U.S. Petroleum Strategies in the Decade of the Environment

U.S. Petroleum Strategies in the Decade of the Environment

Bob Williams

PennWell Books

PENNWELL PUBLISHING COMPANY
TULSA, OKLAHOMA

Copyright © 1991 by
PennWell Publishing Company
1421 South Sheridan/P.O. Box 1260
Tulsa, Oklahoma 74101

Library of Congress cataloging in the publication data

Williams, Bob.
    U.S. petroleum strategies in the Decade of the Environment/Bob
Williams.
        p. cm.
    Includes bibliographical references and index.
    ISBN 0-87814-365-3
        1. Petroleum industry and trade—Environmental aspects—United
States.   2. Petroleum industry and trade—United States.   I. Title.
II. Title: US petroleum strategies in the Decade of the Environment.
TD195.P4W55   1991
333.8'23—dc20   90–27360
                                                                  CIP

Printed in the United States of America

1 2 3 4 5 94 93 92 91 90

To:

Heidi and Wes, for their love, support, and patience;

Mutti, for the first of my life; and especially,

Aletha, for the best of my life.

CONTENTS

CONTENTS

foot	ft
mile	mi
barrels	bbl
barrels per hour	B/hr
barrels per day	B/D
million cubic feet per day	MMCFD
square mile	sq mi
per gallon	/gal.
trillion cubic feet per year	TCF/yr
trillion cubic feet	TCF
cubic feet	cu ft
gallon	gal.
inch	in.
meter	m
per hour	/hr
gallons per day	GPD
acre	ac
miles per hour	MPH
gallons per month	gal./month
per year	/yr
gram per mile	g/mi
pound per square inch	psi
grams per gallon	g/gal.
parts per million	ppm
killowatt	kw
killowatt-hour	kw-h
megawatt	MW
gigawatt	GW
deadweight tons	DWT

ADEC	Alaska Department of Environmental Conservation
AGA	American Gas Association
ANWR	Alaska National Wildlife Refuge
API	American Petroleum Institute
AQMP	Air Quality Management Plan
BACT	Best Available Control Technology
CAA	Clean Air Act
CARB	California Air Resources Board
CAWG	Clean Air Working Group
CCC	Calfornia Coastal Commission
CEC	California Energy Commission
CERCLA	Comprehensive Environmental Response, Compensation, and Liability Act (Superfund)
CIPA	California Independent Petroleum Association
CZMA	Coastal Zone Management Act
DOE	Department of Energy
DOG	Division of Oil & Gas
DOI	Department of Interior
EIA	Energy Information Administration
EPA	Environmental Protection Agency
EPI	Environmental Protection Initiative
FAA	Federal Aviation Administration
GOO	Get Oil Out
GPA	Gas Processors Association
HSWA	Hazardous and Solid Wastes Act
INGAA	Interstate Natural Gas Association of America
IPAA	Independent Petroleum Association of America
IRM	Institute for Resource Management
MACT	Maximum Achievable Control Technology
MSS	Minerals Management Service
NAAQS	National Ambient Air Quality Standards
NAM	National Association of Manufacturers
NAS	National Academy of Sciences
NEPA	National Environmental Protection Act
NERC	North American Electric Reliability Council
NES	National Energy Strategy
NOAA	National Oceanic and Atmospheric Administration

ACRONYMS

NOIA	National Ocean Industries Association
NRDC	National Resources Defense Council
NTSB	National Transportation Safety Board
NWF	National Wildlife Federation
OCS	Outer Continental Shelf
OCSLAA	Outer Continental Shelf Lands Administration Act
OPEC	Organization of Petroleum Exporting Countries
OSHA	Occupational Safety and Health Administration
PIRO	Petroleum Industry Response Organization
PMAA	Petroleum Marketers Association of America
PSD	Prevention of Significant Deterioration
PURPA	Public Utilities Regulatory Practices Act
RACT	Reasonably Available Control Technology
RCRA	Resource Conservation and Recovery Act
RRT	Regional Response Team
SARA	Superfund Amendments and Reauthorization Act
SCAG	Southern California Area Governments
Scaqmd	South Coast Air Quality Management District
SLC	State Lands Commission (California)
TAC	Tanker Advisory Center
TAPS	Trans-Alaska Pipeline System
USGS	United States Geological Survey
WSPA	Western States Petroleum Association

PREFACE

As this book was going to press, the world was again gripped in the throes of another crisis in the Middle East. Although Iraq's brutal invasion and takeover of Kuwait is having a dramatic effect on oil markets as of this writing, it could not be foreseen whether the outcome will result in higher oil prices for a fairly long period or a quick return to stability.

Abruptly, the wave of environmentalism sweeping the oil industry has, for the moment, been pushed to the background to make way for heightened fears over energy security and the effect of higher oil prices on the world's economies.

There is no way to safely predict the outcome of today's events. A swift resolution of the crisis could return markets to stability quickly. A protracted or wider struggle in the Persian Gulf could mean higher oil prices for several years.

My own guess is that any oil supply crisis probably will not last beyond spring 1991. Even if it does, it is only a question of time before market forces come into play to resolve any supply/demand crunch. Conservation, fuel-switching, alternate fuels, marginal oil resources—each will contribute to a lessening of demand to bring the market back into balance again, as it did in the 1970s and 1980s. With the rapid evolution of the futures markets' ability to transmit price signals, the rollback of prices as stability returns should occur more quickly than after the earlier crises. Oil prices will come down again, as will fears over energy security.

With that in mind, the petroleum industry should not be lulled into thinking that the crisis of the moment has squelched the grassroots environmental renaissance. Quite to the contrary, the industry should brace for another wave of anti-oil sentiment once the crisis has passed and environmental concerns leap to the forefront again. Public opinion polls of the day show that the American public believes the run-up in gasoline prices immediately after Saddam Hussein's brutal grab for his OPEC neighbor is the result of oil company greed, not the result of market forces. The industry's feeble attempts to educate the public on Gasoline Marketing Economics 101 did little to disabuse the public of that notion; its price freezes and cuts helped, however temporary they may prove.

U.S. industry may benefit in several ways from events in the Middle East. A renewed awareness of energy security will help the push to cut oil use, especially dependence on imported oil. That means greater use of domestic natural gas where feasible, renewed support for efforts to gain access to Alaska

National Wildlife Refuge's (ANWR) Coastal Plain and the Outer Continental Shelf (OCS), and perhaps tax incentives in support of marginal U.S. oil resources.

But don't expect the environmental movement to scurry off into the corners to hide from the glare of an oil price shock. Environmentalists will use the crisis to push for the alternate fuels of their choice, arguing for a phase-out of oil use for U.S. energy security as well as for environmental reasons.

And don't overlook the parochial nature of many environmental concerns. That can be seen in the response of Santa Barbara County to the Department of Energy's unprecedented offer to mediate in the dispute over bringing the long-delayed Point Arguello project onstream off California. Although the project could yield, perhaps in a year, as much as 100,000 B/D of oil supplies at a time of a looming oil supply crisis, the county did not budge on its stand against interim tankering of the oil—sticking to its preference of a feasible pipeline alternative that could be as much as three years away.

That, I think, is a microcosm of the post-Saddam environmental debate for the U.S. oil industry. When oil prices are flirting with $18–$20/bbl again, consumers soon will forget energy security, Persian Gulf crises, and the threat of a United States hurtling toward 70% oil import dependence.

What they will remember is that an industry—dubbed by one P.R. consultant at the time of the crisis as the "Darth Vader" of American business in terms of public image—took advantage of the crisis to "gouge" consumers at the gasoline pump. That is not the reality; it is, however, the perception. Unless the U.S. oil industry does a much better job than it has to date in getting its case before the American public, then its job in formulating strategies in the Decade of the Environment will get even tougher.

Bob Williams
Tulsa, Oklahoma
November 1990

1

THE CATALYST: EXXON VALDEZ SPILL

It is 9:15 P.M., Thursday, March 23, 1989, Valdez harbor, Alaska. After taking on 1,264,155 bbl of North Slope crude oil from the terminal of the Trans-Alaska Pipeline System (TAPS), the Exxon Valdez supertanker departs for refineries at Long Beach and Benicia, California.

The tanker is of recent vintage, touted with the fanfare of press releases and black and white glossies to the media when Exxon Shipping Co. took delivery of it from its South Korean manufacturer in 1986. The vessel has a design draft of 56 ft and can carry 1.6 million bbl of oil but is light-loaded for California ports. A local harbor pilot is aboard.

At 10:53 P.M., the tanker clears Valdez Narrows, changes course to 219 degrees to enter Valdez Arm and is en route to Prince William Sound and the Gulf of Alaska. About a half hour later, the local harbor pilot disembarks as the tanker nears the outbound of the sound's traffic separation lanes.

In another 15 minutes, the Exxon Valdez contacts the Coast Guard to inform the agency it is changing course to port 185 degrees. Ice floes from the Columbia Glacier are drifting too close on the starboard side. The change puts the vessel across the separation zone and the inbound lane.

The weather is moderate for the region at this time of year. North winds are blowing at 10 knots and carrying a slight mix of drizzle, rain, and snow. It is 33°F and visibility is 10 mi. There is no lookout posted at the tanker's bow from 11:15 P.M. to 11:50 P.M.

(Author's note: The accounts in this chapter of the events leading up to and including the Exxon Valdez spill as well as the initial response and cleanup efforts were taken from Coast Guard logs and reports by the National Transportation Safety Board.)

On board the Exxon Valdez, Capt. Joseph Hazelwood and Third Mate Gregory T. Cousins discuss maneuvers to avoid ice before Hazelwood goes below to his quarters. Cousins disengages automatic gyros he had not realized were engaged, returning the vessel to manual steering. Company policy forbids use of gyros except in open water. Cousins does not have a pilot's license. Leaving a vessel in charge of someone without a pilot's license is a violation of Coast Guard regulations.

At 11:55 P.M., the tanker remains on course at 185 degrees inside the inbound traffic lane. Cousins fixes the tanker's position abeam of Busby Island with Bligh Reef ahead and gives an order for 10 degrees right rudder at 11:56 P.M. His intention is to avoid the reef and return the vessel to the southbound shipping lane.

The change doesn't take effect for five to six minutes. It should have occurred in one minute at the most if the order had been carried out. Cousins realizes no significant course change has occurred and orders 10 degrees right rudder again. He can tell by the tanker's position relative to the lights marking the shallow waters around the reef and on Busby Island that the vessel is in serious trouble. He calls Hazelwood on the telephone.

The tanker strikes bottom several times, hitting a rock outcrop on the starboard side as Cousins and Hazelwood are on the phone. The vessel then runs hard aground on the reef, its heading swinging around 90 degrees. The impacts gouge huge holes in the tanker, rupturing eight of its 13 tanks (Fig. 1–1). Hazelwood gives orders to try to free the grounded vessel, but they fail.

Within one hour, more than 150,000 bbl of North Slope crude escapes into the sound. In all, the spill volume will total 258,000 bbl, the worst in North American waters.

At 12:28 A.M., the Exxon Valdez radios the Coast Guard and reports that it is aground and leaking oil. Two hours later, Alyeska Pipeline Service Co. launches an observation tug to the scene without spill containment equipment. The Coast Guard closes the Port of Valdez to all traffic.

The Coast Guard and Alaska Department of Environmental Conservation (ADEC) officers depart Valdez for the spill site at 1 A.M. The arrival of high tide an hour later fails to refloat the tanker. At 2:27 A.M., the merchant vessel Sherikoff reports an oil slick one half mile south of the Exxon Valdez. An hour later, the Coast Guard boards the stricken tanker and reports that an estimated 138,000 bbl of oil is in the water.

Schematic of Exxon Valdez Showing Damaged Tanks

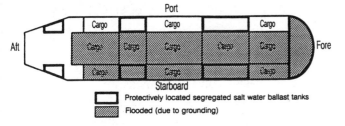

Prince William Sound

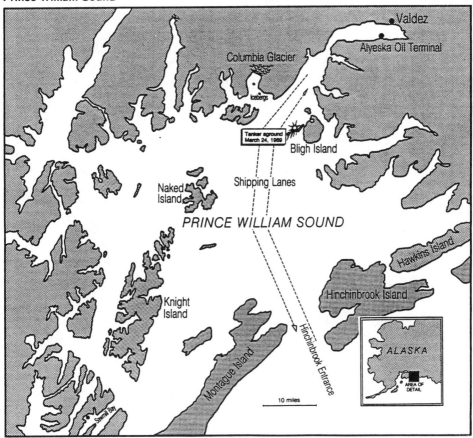

Figure 1–1 Schematic of Exxon Valdez; Prince William Sound.

At 4:14 A.M., the Exxon Baton Rouge is en route to the spill site to begin lightering remaining oil from the Exxon Valdez. Time of arrival is estimated at 11 A.M.

An Alyeska helicopter reports an oil slick 1,000 ft wide and 4–5 mi long heading south at 7:27 A.M.

At 9 A.M., the Coast Guard tests Hazelwood and Cousins for alcohol and/or drug use. Cousins passes. Hazelwood's blood shows an alcohol level of 0.061%. The legal limit for alcohol levels in the bloodstream of someone piloting a vessel is 0.04%.

Alyeska boats depart at 9:50 A.M. with containment and skimming equipment. They estimate arrival time at 1 P.M.

By 11:15 A.M., the rate of oil discharge into the water has slowed.

At noon, the regional oil spill response team (RRT) is talking about the possibilities of using dispersants or in situ burning to contain the spill.

The Coast Guard opens its pollution fund with $25,000 for staging its strike team equipment at 2 P.M. A half hour later, the first Alyeska equipment arrives at the tanker. The Coast Guard gives permission for Alyeska to test dispersants on the leading edge of the oil sheen.

Gov. Steve Cowper arrives at the Exxon Valdez at 5 P.M. The first dispersant test is conducted unsuccessfully an hour later. The governor leaves about 7 P.M.

At 8:10 P.M., the Exxon Baton Rouge arrives alongside the Valdez. In about two hours the first hose is connected between the two tankers.

At 7:36 A.M., March 25, lightering operations begin. Divers survey the Exxon Valdez hull and report massive damage. Lightering continues at a rate of 10,000–12,000 bbl/hr, and seawater is pumped into the Exxon Valdez ballast tanks to keep the vessel stable.

Meantime, Alyeska has cut flow on the TAPS line to 768,000 B/D because of the reduced tanker traffic. It has a week of storage capacity remaining.

The RRT team meets at 9:45 A.M. Exxon announces it has assumed financial responsibility and coordination of the oil spill cleanup. The Coast Guard Pacific region strike team arrives and delivers equipment to the Exxon Valdez deck.

At 11 A.M., the Exxon Valdez is surrounded by containment boom.

The delays in deploying containment boom stem from lack of a barge. Alyeska's barge was in dry dock for repairs.

By this time, 5 skimmers have been deployed in the sound, deploying about

3 mi of boom. By the end of the day, 5 more skimmers deploy another 5 mi of boom.

The spill has spread to an area of 10 sq mi.

At 8 P.M., March 25, a Coast Guard C-130 aircraft dumps Exxon Chemical Co. dispersant on the oil slick. The effort fails. Exxon completes first mapping of the area to determine where the oil spill has spread and which beaches have been affected.

Meanwhile, a test burn is conducted near Goose Island. It consumes about 360 bbl of crude, leaving behind a 100 sq ft area of tarlike substance.

By Sunday morning, Exxon has put to work 350 persons on spill cleanup and has another 200 on standby. The Coast Guard, Alaskan environmental officials, and Exxon begin mapping out a strategy to protect sensitive herring roe areas at the north end of Bligh Island.

Early Sunday afternoon, the weather begins to worsen. Winds pick up speed. Investigators from the National Transportation Safety Board (NTSB) come aboard the Exxon Valdez.

The Coast Guard on scene coordinator (OSC) approves use of dispersants after a second demonstration via C-130 proves successful at about 4 P.M. However, the dispersants to be used in the full deployment have not been loaded onto the C-130. It takes four hours to load the dispersants. By this time, it is too dark to deploy the dispersants, so the drop is scheduled for Monday morning.

Meanwhile Sunday evening, Gov. Cowper has declared a state of emergency. Divers begin videotaping the Exxon Valdez hull. Wind speeds are approaching 70 mph, creating 6-ft waves. Another scheduled in situ burn is canceled as a result.

Monday morning, March 27, winds topping 70 MPH overnight have driven the oil slick to the southwestern corner of the sound about 40 mi. The high winds ground all aircraft for an in situ burn or dispersant drop. Another attempt to spray dispersants on the spill is attempted late in the day, but by now the spill is so widespread it is too late for the dispersants to be effective. The spill now appears to be out of control.

On Tuesday, March 28, the spill has spread to cover more than 100 sq mi. President George Bush sends administration officials to the site to investigate Exxon's handling of the cleanup, hinting of a possible federal takeover of the cleanup operation. Alaskan agencies then take over management of the cleanup from Exxon. Containment efforts are dropped in favor of redeploying booms to protect fisheries and other sensitive areas.

Market Response

. . . .

West Coast refiners responded quickly to the disaster, scurrying to locate other crude supplies after the shutdown of the Valdez terminal raised fears of refinery throughput curtailments and possible gasoline shortages.

Those fears in turn spiked up gasoline futures prices. West Coast motorists began topping tanks. Spot shortages of gasoline developed. West Coast wholesale gasoline prices rose by about 7¢/gal. About three-fourths of North Slope crude production of about 2 million B/D in early 1989 moved to refineries on the West Coast, the rest mostly to refineries on the Gulf Coast. That dependence could have easily translated into widespread shortages in the West.

Chevron relies on Alaskan crude for about one-third of its West Coast refining needs. The company came within 24 hours of having to curtail throughput at its plants, which would have triggered widespread product shortages on the West Coast.

Those fears dissipated when the Coast Guard reopened Valdez terminal to limited tanker traffic March 28. Tankers were limited to daylight hours, meaning at most four vessels could load and leave before nightfall.

Gov. Cowper demanded traffic be restricted in the sound until Alyeska provided an oil spill contingency plan capable of containing a spill comparable to the Exxon Valdez spill. Without those assurances, the governor would close TAPS.

Meanwhile, West Coast refiners were still grappling with the reduction in Alaskan oil supplies. BP America on March 30 declared force majeure on April crude contract sales, expecting to deliver at most 80% of contract volumes totaling about 800,000 B/D. Exxon, with a 20% share of TAPS throughput, followed suit with force majeure for April deliveries.

ARCO, with a major turnaround at one of its refineries under way, was not as hard pressed to meet contract deliveries. It didn't declare force majeure but warned customers of delays in deliveries. ARCO cut throughput by 70,000 B/D at its refinery at Cherry Point, Washington. There was enough product in storage, however, to meet demand.

Chevron shut down its 18,000 B/D refinery at Kenai, Alaska, which produces mostly diesel and jet fuel. It found an easy solution in diverting crude shipments to its Richmond, California, refinery and then shipping the products to Kenai. Chevron bought about 1 million bbl of crude from Sumatra and 750,000 bbl of crude from Ecuador to make up the lost Alaskan supplies.

In all, North Slope producers lost about 6.5 million bbl of crude sales during the terminal shutdown. Refiners drew down stocks and moved in other supplies from the Gulf Coast or via imports.

Spot markets showed the possible market repercussions. On the West Coast, gasoline prices soared 20¢/gal. in the days following the spill. The 7¢/gal. increase in dealer tank wagon prices came on top of a 15¢/gal. rise since the first of the year. The earlier rise was led by other factors, such as new air quality rules in the U.S. Northeast expected to cause product shortfalls in the spring and summer and the reflection of crude oil price hikes at the end of 1988 owing to OPEC production discipline.

Political Fallout

. . . .

The political response to the Exxon Valdez spill was swift and clear: Big Oil had once again raped the public trust. There were howls of indignation and calls for retribution in the halls of Congress. Environmental organizations, without much of a rallying cry in the years since Ronald Reagan's Secretary of Interior James Watt left the scene, moved quickly to make political capital out of the tragedy. Although their top priority in the past two years had been the fight over leasing and drilling on the Coastal Plain of ANWR, it had not gained much of a foothold with the public at large over the issue.

After all, the reasoning went, the industry has had a good environmental record in Alaska, the doomsday warnings from environmentalists during the building of TAPS notwithstanding. In the 11-year history of the Valdez terminal, 9,000 tanker trips carrying almost 7 billion bbl of oil had been made without a major spill off Alaska. A leaked draft study by the EPA purporting to show a history of systematic damage by North Slope producers was handily discredited by industry and federal witnesses in 1988. The outlook for some sort of ANWR leasing bill making headway in that Congress had been improving.

Prospects for new leasing off other areas, such as California and Florida, were dimmer. President Bush, anxious to prove himself an environmentalist, probably saw giving up leasing off California and Florida as a political tradeoff on ANWR. Although environmentalists topped their agenda with ANWR, they did not have the support in Alaska that could provide them the leverage they needed to make it a political threat to the Bush administration.

Also weighing against the environmentalist lobby on ANWR was the wide-spread recognition that the remote arctic desert offered the United States its single best chance at finding another Prudhoe Bay. With Lower 48 production falling sharply, oil imports soaring, and the supergiant North Slope field entering into decline, there loomed the specter of rising oil prices and 60–70% import dependence as a backdrop to the 1992 elections. A political tradeoff was definitely needed. On federal lands access for the oil and gas industry, it seemed, the Bush administration had put all its eggs in one basket: ANWR's Coastal Plain.

Environmental lobbyists wasted no time in capitalizing on the incident with a barrage of rhetoric that amounted to a single statement, with subtext: *We told you so, and the oil industry lied to you.* Environmentalists now had a rallying cry and, because of industry's widely unpopular image in the 1970s and early 1980s, an ideal villain.

The hyperbole reached stratospheric heights. The judge that set bail of $1 million for Exxon Valdez master Capt. Joe Hazelwood justified the absurd amount by declaring the spill damage as the worst example of man-made destruction since Hiroshima. Equating an oil spill, even in the awesome, pristine beauty of Alaska, with the nuclear devastation of a city with its tens of thousands dead and legacy of disease and deformity, reflects the sensibility of environmental extremism today.

With the political climate as fevered as it was, it quickly became imperative that industry and government's early response to the disaster be as effective as possible to minimize the political fallout. That was not to be.

Cleanup Begins

· ·

Cleanup efforts in Prince William Sound were complicated from the beginning by a combination of factors involving weather, logistics, confusion over authority and responsibility, and inadequate equipment.

The initial response at the Port of Valdez under Alyeska's control was hindered by damaged equipment and shortages of personnel due to the Easter weekend holiday. Alyeska's contingency plan calls for the company to provide an initial response at the affected vessel within five hours of first notification. It missed that goal by more than *seven* hours.

The only containment barge was tied up and stripped for repairs at Valdez terminal. Although not certified by the Coast Guard to receive oil, the barge was capable of carrying recovery bladders. Alaska's contingency plan calls for Alyeska to notify the state when response equipment is taken out of service. Because the company was satisfied that the barge was seaworthy without repairs, it had not notified the state.

By the time Alyeska called crane riggers to load pollution gear onto the barge, three hours after being notified, about 138,100 bbl of oil had discharged from the Exxon Valdez tanks. The barge carried two 1,000 gal. bladders and 8,000 ft of containment boom and towed 2 skimmers.

By then, the oil slick had grown 1,000 ft wide and 4–5 mi long.

The next day, March 25, at 7:30 P.M., cleanup crews estimated they had recovered 1,200 bbl of oil.

Meanwhile, wildlife rescue efforts had begun. A representative of the International Bird Rescue and Research Center arrived in Valdez to set up facilities for treating oiled animals. Exxon set up the bird rescue operation.

By the end of the fourth day of the emergency, March 27, skimmers had recovered only 3,000 bbl of oil. Fierce winds overnight had driven the oil spill almost 40 mi, coating beaches at Little Smith, Naked, and Knight islands. Eleanor Island was heavily oiled. The weather forced skimmer systems, booms, and other equipment to sheltered waters for protection. Mechanical recovery operations came to a halt. Although permission was given for dispersants to be used, all aircraft were grounded.

On March 28, Exxon asked the RRT for permission to use dispersants and try in situ burning around Eleanor Island. The RRT deemed dispersants inappropriate in the area. About an hour later, ADEC approved in situ burning for the Eleanor Island area.

Management of the spill response was reorganized March 28 as a steering committee of Exxon, the Coast Guard, and ADEC. A major cleanup effort was mobilized to protect critical fisheries in Eshamay Bay, Main Bay, Port San Juan, and Esther Bay.

The Coast Guard OSC authorized use of dispersants in response to a minor discharge of oil from the Exxon Valdez. Exxon chartered a C-130 to drop dispersants in areas of heavy oil concentrations near the Exxon Valdez and eastern end of South Island, reporting excellent results.

With the Port of Valdez open to vessel traffic again, only day transits were allowed, along with two-tug escorts to or from Bligh Reef. A 1,000-yd safety zone encircled the Exxon Valdez.

At this point, Exxon's cleanup force had grown to include 71 vessels, 34,000 ft of boom, 7 skimmers, and 340 personnel. The Coast Guard complement included a high endurance cutter, a helicopter, a buoy tender, a 32-ft boat, an Aireye Falcon plane, a C-130 transport plane, 5 air deployable antipollution transfer systems, and a 16-member strike team with an open water oil containment and recovery system. Also on hand was a NOAA helicopter and about 45 representatives of other federal agencies. That didn't include the large contingent of people from Alaska's private and public sectors.

In fact, much of the early response success in protecting sensitive fish hatcheries and spawning grounds can be credited to the local communities and fishermen who put together an impromptu flotilla to protect sensitive habitats with often makeshift means.

On March 29, Transportation Sec. Samuel Skinner, EPA Administrator William Reilly, Coast Guard Commandant Adm. Paul Yost, Alaska Sen. Frank Murkowski, and Congressional staffers reviewed cleanup and oil transfer operations and met the OSC in Valdez. The Exxon San Francisco tanker left to take over lightering of the Exxon Valdez.

The following day, Exxon and the Coast Guard established three separate beach cleanup work groups: one to rank most critically affected areas for cleanup, another to decide the best cleanup methods, and the third to make final assessments of cleanup work.

By this time, the oil slick had passed Montague and Latouche islands and was moving west into the Gulf of Alaska. NOAA estimated that 30–40% of the crude spilled had evaporated. Exxon by then had recovered only 7,537 bbl of oil.

The Exxon San Francisco arrived to take over oil transfer operations. The Exxon Baton Rouge departed with about 447,000 bbl of oil from the Exxon Valdez, which held a remaining 668,000 bbl of oil.

On the eighth day of the crisis, overflights showed that the oil slick appeared to be turning back on itself. The oil began to emulsify with the water, increasing the volume of liquid to be recovered. The cleanup teams switched to rope mop skimmers from weir type units.

An early Interior survey of Green Island indicated 1,000 birds had been oiled. Three dead sea otters were recovered from the sea. Another 10 otters were received at the wildlife recovery center.

A Coast Guard cutter transferred almost 80,000 ft of sorbent boom to contractor vessels in the southern portion of the sound.

There were more reports of heavily weathered oil on April 1. Exxon by then had deployed more than 84,000 ft of boom, with 6,000 ft around the Exxon Valdez. State officials began taking water samples in hatcheries and spawning areas. There were 28 birds and 12 otters being treated at the wildlife recovery centers.

By April 2, the leading edge of the spill was about 9 mi south of Resurrection Bay and moving southwest. Exxon's team totaled 160, including spill experts from the United States, Canada, and the United Kingdom. It hired more than 350 additional cleanup workers and employed almost 100 vessels to participate in the cleanup operations.

The federal employee population on the scene had swollen to almost 500 from a dozen agencies, primarily the Coast Guard.

With the Exxon San Francisco loaded to capacity, the total lightered from the Exxon Valdez reached 943,000 bbl of oil. The Exxon Baytown arrived alongside to continue oil transfer.

Exxon tried applying dispersants to a slick south of Point Erlington, but they were ineffective on the main body of the spill.

The wildlife recovery centers had treated 150 birds and 30 otters, shipping the treated otters to aquariums around the country.

Exxon received offers of equipment and help from around the world. France sent two belt skimmers, and the USSR loaned a skimmer/dredger. The Oil Spill Service Center in Southampton, England, sent five staffers to Alaska accompanied by 58 tons of equipment including oil skimmers and booms. The U.S. Navy loaned 13 skimming systems and the Coast Guard two skimming barrier systems.

On April 3, Alaska's Department of Fish & Game canceled all herring fishing in Prince William Sound because of the damage to spawning areas. The next day, the Exxon Baytown completed its lightering of crude from the Exxon Valdez.

More than 66,000 ft of boom—65% of the total—had been deployed in Sawmill Bay by April 5. Taking so much boom out of spill containment was deemed necessary as a defensive measure to protect hatcheries.

The U.S. Air Force airlifted in Navy, Coast Guard, and Exxon skimmer boats, dispersant, mooring systems, boom vans, barrier material, and assorted vehicles from California, Oregon, Texas, Virginia, Denmark, and Finland.

Fishermen in Sawmill Bay began to grow confident that hatcheries would be protected. Beach cleanup crews began mopping up oil in tidal pools. Skimming rates were getting worse because of the oil weathering. Recovery of oil became extremely difficult as it began to form into mousse.

At 10:35 A.M., the Exxon Valdez was refloated and held position on Bligh Reef. About 16,445 bbl of oil remained on board. Exxon began towing the stricken tanker to anchorage in Outside Bay near Naked Island. The company placed two skimmers and a vacuum truck aboard and backed them up with workboats and a standby aircraft laden with dispersant.

The U.S.–Canadian joint contingency plan was activated on April 6 to speed delivery of more cleanup equipment and operators. Another 21 skimmers were en route. A thinning oil slick continued to move into the Gulf of Alaska, and ribbons of oil reached the entrance of Resurrection Bay, putting affected areas into the jurisdiction of the Anchorage OSC.

Exxon encountered problems positioning skimmers in areas of heavy oil concentration. Larger skimmers could not reach some shorelines where there were emulsified patches of oil.

On April 7, President Bush ordered the Defense Department to make available all facilities, equipment, and personnel that could be effectively utilized in assisting the cleanup. The military was to provide personnel for direct cleanup activities as well as logistics assistance, but intensive planning and appropriate cleanup training were to be carried out before deploying ground units.

Despite the environmental concerns the spill had stirred up, Bush said, the oil spill was "an aberration and should not affect U.S. oil exploration."

Even with the Coast Guard taking charge of the cleanup operation, the President remained opposed to federalizing the effort. Coast Guard Adm. Yost warned against federalization, noting that such a move might allow Exxon to close its checkbook and force the government to assume cleanup costs.

Meanwhile, about half of the sea otters at the rehabilitation centers had died. Cries of environmental devastation and criticism of the way the spill was being handled mounted daily. The increasing death toll of birds and marine mammals was issued by the animal rescue centers with a numbing regularity.

The Center for Marine Conservation, two weeks after the spill, called on President Bush to provide more rescue centers, or thousands more birds and mammals would die needlessly from the spill's effects. The conservation group said it took a team of five people two hours to clean one oiled sea otter.

Removing oil from the feathers of a bird required one worker for an hour and 150 gal. of heated water. At the time, there were only two emergency centers and one veterinarian to handle the incoming wildlife. The group estimated that about 80% of the birds being brought in were dying because of a lack of facilities. The group figured it would take at least 10 centers, each supervised by four marine mammal experts, to save most of the wildlife that was endangered.

Medical reports were being issued daily for some of the stricken animals, accompanied by gruesomely clinical details of the animals' death throes. Ingesting crude oil caused liver and kidney damage to the sea otters. Television reports showed the rheumy red eyes of otters with toxic crude levels in their systems as the animals inexorably weakened. In all, about 300 dead birds and 76 sea otters were collected by the time the new Valdez rehabilitation center started up April 7.

The spill by then covered about 2,600 sq mi, and shifting winds had nudged the slick towards the Kodiak Island area, the No. 1 U.S. fishing port. Kodiak is about 300 mi from Bligh Reef. Its most recent annual catch was valued at more than $160 million.

Problems continued to plague the cleanup effort. A shipment of U.S. Navy containment booms proved to be only 8,000 ft instead of the 16,000 ft promised and the wrong size as well.

Increasing viscosity of the oil slashed oil recovery rates to 200 B/D in Coast Guard skimmers rated at five times that level.

Interior, setting up a wildlife collection station in Whittier, Alaska, put the death toll at 529 birds and 94 sea otters on April 8.

By this time Exxon had recovered 17,000 bbl of oil. Of the initial volume of spilled oil (then estimated at 240,000 bbl), the remaining volumes were estimated at 77,000 bbl evaporated, 11,000 bbl dispersed, and 45,000 bbl each in Prince William Sound, in the Gulf of Alaska, and on the beaches.

As of April 9, the 17th day of the crisis, the spill seemed to be stabilizing, with the leading edge not advancing for two days. The oil spill response armada continued to grow, joined now by a 1,000-person floating hotel, 5 waste oil barges, 5 waste oil doughnuts, and 100 small skiffs.

Stormy weather began to break up the spill on April 10, but it also hampered cleanup operations. Coast Guard efforts to deploy an 84-in. boom at Seward failed because of high seas. Overflights were hindered by poor visibility and high winds reaching gale force the next day.

Fishing vessels dragging herring nets formed a mobile response unit to break up oil patches at Cape Resurrection. Earlier tests with the nets proved successful, breaking the oil into small droplets that sink.

At that point, there were about 200,000 ft of boom deployed in the sound, but 39 skimming operations were shut down for two days because of the bad weather. High seas and winds began alternately mixing and dispersing the bigger concentrations of oil in the sound's open waters.

Adm. Yost met with Exxon officials to set cleanup priorities on April 13. Exxon, obliged to submit a beach cleanup work plan, temporarily suspended shoreline cleanup pending submission and approval of the plan. The Coast Guard, by presidential order, was now officially in charge of the cleanup.

Cleaning Up

Entering the third week of the aftermath of the worst oil spill in North American waters, Exxon and the Coast Guard began to map out a strategy for cleaning oiled beaches amid a welter of controversy.

Progress reports would fill the nation's newspapers and air waves on a daily basis for weeks. It was easily the No. 1 news story at the outset, and kept environmental concern at the forefront of public consciousness in the spring and summer.

It was a standard joke of the day at the time of early reaction to the spill that souvenir T-shirts related to the disaster were available before Exxon could get a containment boom around the stricken tanker.

The bitterness and anger at Exxon and the petroleum industry in general was reflected in the editorial cartoons of the day, some of them scathing indictments of "Big Oil" as inept, arrogant, greedy, and heedless of environmental concerns.

Exxon seemingly did little to disabuse the public of that image, with officers of the company making comments or taking actions that may have seemed acceptable from a business management standpoint, but were disastrous from a public relations standpoint.

Exxon Chairman Lawrence G. Rawl's decision not to be on the scene in the early days of the crisis was widely criticized. Although he contended he could not have contributed to managing the crisis by being there in person, he later admitted the decision was a mistake.

14

Another gaffe occurred when a senior Exxon official noted that Exxon planned to reimburse federal and state agencies for cleanup costs, but expected to deduct from its taxes much of the cleanup cost as a cost of doing business. That elicited a howl of indignation. Sen. Harry Reid, a Nevada Democrat, quickly introduced a bill into Congress to prevent companies from deducting spill cleanup costs from their federal taxable income if the cleanup is not done to EPA's satisfaction.

The political repercussions continued to mount, even beyond U.S. borders. British Columbia extended through 1994 a ban on offshore drilling.

One of the early political victims of the Exxon Valdez spill was the effort to open ANWR's Coastal Plain to oil and gas leasing. Industry explorationists consider the ANWR Coastal Plain the most prospective area for finding giant oil and gas fields in North America, but environmentalist groups consider ANWR an untouchable wilderness and keeping it untouched their number one priority. Under current federal law, it would take an act of Congress to allow leasing and drilling on the ANWR Coastal Plain.

The industry had been gaining some ground on getting ANWR legislation shepherded through Congress, but the Exxon Valdez spill changed all that. The environmental lobby's surrogates in the Congress lost no time on connecting the spill to ANWR leasing, despite industry's protests that the two issues were not connected.

Support for an ANWR bill all but disappeared. The House bill on ANWR leasing was shelved indefinitely. Even a friend of the industry such as Louisiana Democrat Sen. Bennett Johnston abandoned his ANWR leasing bill, recognizing the political futility of the effort.

Committees and subcommittees in the Senate and House scrambled to take testimony on oil spill response and cleanup, oil transportation safety, oil spill liability, offshore leasing and drilling, federal lands access, and other issues. Some of the effort spawned legislation, some was mere posturing for constituents.

Responses likely to have a more far-reaching effect were those by Alyeska, the NTSB, and the API.

Alyeska moved to improve its oil spill contingency plan, partly at the behest of Alaska. The company put backup equipment in stock in Valdez to replace that deployed for the Valdez spill. It began a double tug escort service for all tankers outbound from Valdez. Alyeska also complied with a state order to restock the Valdez terminal with skimmers and surround all tankers with containment boom during crude loading operations.

15

There were now 12 persons plus supervisors on watch around the clock at the Valdez terminal whose sole function was to respond to a spill.

Alyeska began a month-long study of its response to the spill and promised to quickly put into place recommendations from owner companies based on study results.

API endorsed Alyeska's plans and at the same time named a panel of top industry executives headed by Mobil Chairman Allen E. Murray to a task force to review the disaster and report in three months.

The API task force studies ship manning levels, facilities and personnel available in areas of spill cleanups, readiness of personnel and facilities, adequacy of contingency plans, and frequency of exercising contingency plans.

Random drug and alcohol testing of seamen began on most tankers carrying North Slope crude.

The NTSB's final report was issued in midsummer 1990 and offered plenty of blame to go around: Exxon, for overworking tanker crews; Alyeska, for not providing sufficient early response capability; the state, for not providing adequate pilot service; and the Coast Guard, for not maintaining adequate radar.

The federal role in the spill response and cleanup continued to grow. President Bush placed Transportation Secretary Skinner over the operation, granting him authority to coordinate the actions of the 14 federal agencies involved in the cleanup effort. In charge of operations in Prince William Sound was Admiral Yost, working with Exxon and Governor Cowper.

Bush also ordered EPA Administrator Reilly to coordinate efforts aimed at restoring the sound's ecosystem. That was to include replacing wildlife and fish stocks and restoring habitats.

As of April 12, Exxon had contracted for the cleanup campaign 210 vessels, 25 aircraft, and 1,300 workers. The company had 41 skimmers working and deployed 283,000 ft of boom.

The federal contingent as of April 12 included: From the Coast Guard, 4 cutters, 4 buoy tenders, 6 small boats, 3 fixed wing aircraft, 6 helicopters, a 20-person strike team with 2 open water oil containment and recovery systems, 6 air-deployable antipollution transfer systems, 11,200 ft of boom, and 208 personnel; from the Navy, 20 skimmers, 2 barrier skimming systems, 10 tow boats, 20 mooring systems, 11 support vans, and 94 personnel; from the Forest Service, 1 helicopter and 30 personnel; from NOAA, 1 helicopter, 4 data buoys, and 22 personnel; from Interior, 26 personnel; from EPA and the Federal Aviation Administration, 7 personnel each, and 93 National Guardsmen.

Exxon began tallying some of the costs involved in the cleanup, spending millions of dollars per day. Early in the crisis, Rawl predicted that cleanup costs would far outstrip the $100 million in damages that Exxon is insured for through TAPS. The federal government's emergency oil spill cleanup fund is about $3 million.

Estimated costs for oil recovery typically average $12–25/gal, according to Golob's Oil Spill Bulletin.

The loss to Alaska's fishing industry because of the spill was put at $150 million the year of the spill alone. That figure could rise in coming years if there is a reduction in fishing from damage to hatcheries.

Exxon established a claims office in Valdez and set aside funds in local banks to provide advance payments to those in need while their claims were being put together.

The federal pressure on Exxon accelerated. Admiral Yost met with top Exxon officials on April 14 and presented them with a list of 50 beaches needing cleanup. He asked Exxon to provide more personnel within 10 days.

Exxon's shoreline cleanup committee approved use of a wash-vacuum oil-cleaning system (Vikovak) on the eastern shore of Smith Island. Test cleaning using hot, cold, and high-pressure water with Vikovak was approved for the northern end of Smith Island. There was some concern expressed about the cleaning techniques' effect on the ecosystem. Workers were instructed to avoid using high-pressure water or steam where there were invertebrates and seaweed.

Vice Adm. Bill Robbins took over as federal OSC on April 15. Admiral Yost met with Exxon, ADEC, and fishermen as Exxon presented its shoreline cleanup plan. Yost approved the cleanup plan and urged all deliberate speed.

Exxon set a target date of September 15 to complete cleanup of Prince William Sound. The company estimated that it had 305 mi of shoreline to clean in the sound (Fig. 1–2). Of that total, 54 mi were heavily oiled, 85 mi moderately oiled, and 165 mi lightly oiled.

Exxon estimated that of the original 240,000 bbl of oil spilled, 84,000 bbl evaporated, 22,000 bbl were recovered, 12,000 bbl biodegraded in the water, and 12,000 bbl dispersed.

Of the remaining oil volumes, Exxon estimated 30,000 bbl remaining in open waters of the sound, 30,000 bbl on the sound shorelines, 40,000 bbl in open waters of the gulf, and 10,000 bbl on gulf shorelines.

The plan called for beaches to be cleaned by hosing them off with low-pressure, cold water and flushing the oil into the water to be skimmed off.

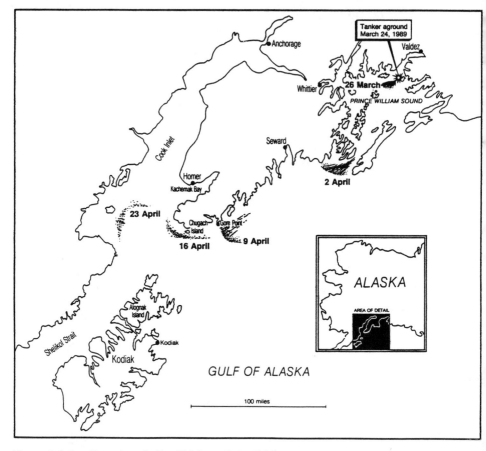

Figure 1–2 Leading edge of oil spill (through April 23).

Another proposed technique, using 110°F water, high-pressure washing, tested on Smith Island, would be more productive, but was itself damaging to micro-organisms and small marine life.

Exxon figured harsh weather would halt operations only about 5% of the time in June and July, 20% in August, and as much as 80% in September and October with the arrival of the storm season.

The company estimated forces directly involved in the cleanup in Prince William Sound alone would peak at about 1,250 persons, 11 barges, 28 landing craft, 150 utility vehicles, and 6 Vikovak units. Support operations would peak at about 413 persons and 140 vessels. In all, Exxon expected to hire about 4,000 persons for the duration of the cleanup.

Exxon also submitted a separate, supplemental cleanup plan for areas outside Prince William Sound. It set up three task forces for cleaning oiled areas in the Gulf of Alaska, Kenai Peninsula, and the Shelikof Straits. These areas are farther from the origin of the spill, and the effects of the spill were less damaging. Much of the cleanup along beaches in these areas had to be done by hand, with workers picking up tar balls and oiled kelp and seaweed and other debris with rakes, shovels, and buckets and putting the debris into plastic bags or waste baskets.

Exxon based the supplemental task forces in Seward, Homer, and Kodiak. It projected the workers involved in cleanup in the three task forces to peak at 582 and targeted a September 1 completion date.

Although the Coast Guard approved Exxon's cleanup plans, it did so with reservations that the company could get the job done in the time allotted.

The clamor over the spill and cleanup never seemed to abate, and Exxon's cost seemed to have no ceiling.

In addition to defending itself over the spill's cause and response, Exxon found itself defending the industry for the spikes in gasoline prices after the spill.

Ohio Democratic Sen. Howard Metzenbaum, long a gadfly to the petroleum industry, claimed the gasoline price increases were the fastest in history and accused the industry of taking advantage of the disaster to gouge hapless motorists. Television newsmen interviewed motorists at gas pumps, eliciting citizens' speculation over price-fixing conspiracies and claims of ripoff amid environmental rape.

However, a look at the nature of the market when the accident occurred shows that gasoline prices were primed for increases in any event. The Exxon Valdez spill only hastened their arrival.

Crude prices had jumped by $6/bbl as a result of strong demand and a successful quota agreement among members of the Organization of Petroleum Exporting Countries the previous fall. The downstream market simply had not fully reflected the crude price increase yet. There is often a time lag—sometimes several weeks—before a crude price increase can be felt at the pump. And if the demand is not there downstream to absorb a hike at the pump, it is only the refiner who gets squeezed. That was the case for much of the 1980s, as skyrocketing prices eventually led to sluggish demand and gasoline price competition and market share battles. But lower gasoline prices following the crude price collapse in 1986 sharply boosted demand, especially for premium, high-octane gasolines. It was the industry's scrambling to meet the demand for higher octane gasolines that revved refineries up to peak utilization rates that sometimes effectively reached 100% and helped make the supply situation tight.

In the weeks preceding the spill, refiners' runs to crude stills, imports of crude and products, stocks, and production of gasoline were the lowest of the year. Refiners were keeping stocks low, in addition to the usual low wintertime demand, because of uncertainty over new rules affecting the volatility of gasoline that the EPA was likely to implement for the summer. Refiners were already trying to meet the reduced volatility limits, and additional costs related to those efforts were starting to show up at the pump. Several refineries were in turnaround in the period to accommodate the fuel specification changes, adding to the supply tightness. In all, the market factors added perhaps 15¢/gal. from the first of the year to mid-March.

The spill's repercussions were affecting upstream areas other than prospects for ANWR leasing and elsewhere on the Outer Continental Shelf. Alaska's Department of Natural Resources' Division of Oil and Gas (DOG) postponed a final decision on scheduling a sale of oil and gas leases in the Cook Inlet, citing a need to give more time for public comment in light of the oil spill. Later, the DOG would suspend the leasing program altogether because the state legislature would not provide funds to sustain it—partly for political reasons stemming from the spill, partly because of Alaska's looming fiscal crisis. The program was resumed in 1990.

Later in May, the Interior Department would delay an environmental impact statement for an August 1990 Gulf of Alaska/Cook Inlet oil and gas lease sale. Interior Sec. Manuel Lujan said that because the Sale 114 area was affected directly by the spill, the EIS would be delayed pending more information about the spill's effects and its relationship to the sale. The document was to have been released in September 1989.

Current offshore drilling and production operations were not about to escape the tightening regulatory vise triggered by the oil spill. Minerals Management Service told operators of offshore drilling rigs and production platforms to expect more surprise inspections.

Alyeska stepped up efforts to enhance its capability to respond to an oil spill in Prince William Sound in May. It acquired three large emergency response vessels to escort outbound vessels in the sound. The escort vessels are equipped to immediately deploy containment booms and begin oil recovery within minutes of a spill. Alyeska also accelerated drug and alcohol testing for all tanker crews, improved navigation aids and communications in the sound, established a 24-hour oil spill response team at the Valdez terminal, resupplied the terminal with oil spill response equipment, and ordered more equipment.

By the start of May, Exxon had about 2,000 people and 370 boats involved in the cleanup. There were about 500 people working directly on shoreline cleanup, the rest in support work. The work proved tedious, backbreaking, and costly. Cleanup workers were paid more than $16/hr plus overtime, most working 14-hour shifts for two weeks at a time. Workdays were sometimes 18 hours or more. Exxon offered Alaskan fisherman $5/gal. of oil brought in bait buckets for collection at barges. Late in the summer, Exxon paid perhaps $100 million/month, for workers and boats alone.

The cleanup work also became extremely frustrating. Because of the porousness of the gravel beaches, reoiling became a common problem. In some places oil settled as much as 1 m deep. As waves washed the shores, they pulled out ribbons of oil that returned with the next rush of waves to reoil the beaches. Some of the scavenged oil also stuck to mats of kelp and seaweed, trailing them as drifting stringers to oil other beaches or worsen the mousse effect.

Workers struggled for days, even weeks, to clean some areas, only to find them stained with crude again by another fugitive slick. Even long after high tide, crude would come percolating up from the depths of some rocky shores.

Exxon modified machines intended for pouring cement in multi-story buildings and irrigating farmland for washing heavily oiled beaches and mounted them on maxibarges. For treating rocky cliff faces and other inaccessible areas, it mounted fireboat nozzles on the maxibarges. Combined, the water delivery systems had a capacity of more than 200 million GPD of water, enough to serve a city of 1 million people. Exxon specially built 1,000 industrial grade water heaters to provide almost 17 million GPD of warm water.

Cleanup became an especially tricky business when the weather got rough. The cranelike omnibooms had to remain close to shore, and when the waves got high, that set up a rocking motion that sometimes crashed the devices into the beach—not to mention the threat to workers nearby.

Exxon also tested one of its chemical products, Corexit, for use as a cleansing agent on several shoreline areas in the sound. Corexit, a kerosine based solvent mixed with detergent, normally is used to clean spills in open seas. There were some concerns about the toxic nature of the chemical, but Alaska agreed to expedite testing.

The cleanup was slowed further—and made still more costly—by disputes over disposal of oily waste. The ADEC, citing toxic air pollution that would result, denied Exxon permits to burn the oily debris in the state. Exxon instead had to ship the waste to a toxic waste dump in Oregon.

There were also disputes about the dangers of the most effective cleaning methods. For some of the toughest cleanup problems, involving weathered oil that proved intractable to cold water flushing, the use of steam or hot water was thought to do more damage than the oil itself.

Cleanup in open water proved just as stubborn. The Soviet oil skimmer Vaidogubsky arrived in the sound from Sakhalin Island in late April to help with cleanup operations. The 425–ft vessel was only able to recover 60 bbl of oil in 2–3 days because of the kelp-matted oil mousse. It was subsequently modified to pick up the heavy, weathered crude.

The death toll in midsummer approached 28,000 birds, almost 1,000 sea otters, and perhaps several hundred seals. The death tolls were misleading, however, because they represented only those animals recovered. That could have been half or less than half of the actual number killed (Fig. 1–3).

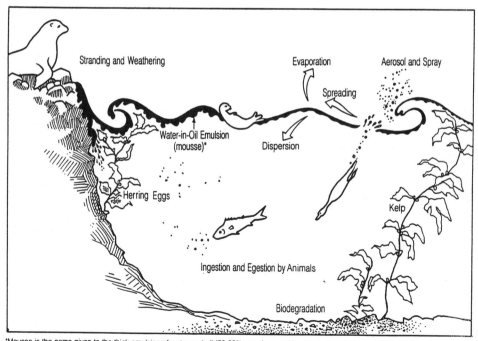

*Mousse is the name given to the thick emulsion of water and oil (50-60% water) caused by the wind and waves.

Figure 1–3 Representation of oil behavior in Prince William Sound.

Commercial fishing in Prince William Sound and the Gulf of Alaska continued to be disrupted in the summer. Most of the fisheries in the sound, Cook Inlet and along Kodiak Island and the Alaska Peninsula were closed. Although there were no reports of contaminated fish, Alaska fishing authorities promptly closed down fisheries or delayed seasons with sighting of oil.

Legal Issues

The Exxon Valdez oil spill might in another context be called the Great Lawyer Full Employment Act of 1989. In the first month of the spill alone, Exxon was hit with 31 lawsuits and claims ranging from $500 to $4 million.

The spill even spawned a newsletter devoted exclusively to covering legal issues emanating from the accident. *Alaska Oil Spill Reporter,* Seattle, reported that the first pretrial conference related to the Exxon Valdez spill was jammed with 75–87 attorneys, packed so tightly before the bar that it was almost impossible to move in the courtroom. The trial for the class action suits might not occur until 1991, the newsletter reported.

It remains to be seen how much Exxon's voluntary claim program will mitigate some of that ligitation. As of spring 1990, Exxon had disbursed more than $200 million under the program for 1989 income losses of fishermen caused by the spill.

Exxon Valdez skipper Joe Hazelwood, probably the most publicly pilloried naval officer of modern times, came out a victor in his court trial. He had been charged with three second degree felony charges and three misdemeanor charges alleging operating a water vessel while intoxicated, reckless endangerment, and discharge of oil into water. He was convicted of the least serious offense, negligent discharge of oil, a misdemeanor that netted him a maximum sentence of 90 days in jail and a $1,000 fine—suspended on the condition he spend 1,000 hours helping clean up the spill. That conviction was still under appeal at presstime. Even if he wins, Hazelwood's legal problems are far from over. He also is named in many of the suits against Exxon.

Complicating the legal issue for Exxon is Hazelwood's lawyers' contention that Exxon is legally responsible for his actions. Hazelwood was reported to have a history of drinking problems and had gone through company sponsored rehabilitation programs. Exxon was aware of the situation yet retained Hazelwood in its employ without sufficient follow-through on his rehabilitation,

Hazelwood's lawyers contend. Exxon fired Hazelwood after the allegations of intoxication came to light.

Exxon also faces criminal charges as a result of the spill. A federal grand jury in early 1990 indicted Exxon Corp. and Exxon Shipping Co. on felony and misdemeanor charges stemming from the spill. At presstime, the trial date was set for April 10, 1991.

According to the Justice Department, if Exxon is convicted on the five count indictment, the company could face fines totaling more than $700 million.

The government brought charges under the Dangerous Cargo Act and the Ports and Waterways Safety Act for the first time in the case of an oil spill. It cited the extensive environmental damage caused by the Exxon companies' alleged "negligence and unlawful conduct."

One indictment alleges Exxon violated the Dangerous Cargo Act by employing a crew member (Hazelwood) who was not physically or mentally capable of performing his assigned duties. The other alleges Exxon violated the Ports and Waterways Safety Act, which requires the owner or master of a vessel to ensure the wheelhouse is constantly manned by competent personnel.

Alaska's lawsuit against Exxon and the other Alyeska partners ARCO, Amerada Hess Corp., BP America Inc., Mobil Oil Corp., Phillips Petroleum Co., and Unocal Corp. could involve ultimate damages running into the billions of dollars and stretch the length of the 1990s. The state alleges Exxon and Alyeska were negligent in preventing and cleaning up the spill.

Exxon, in turn, has countersued Alaska, claiming the state interfered with its efforts to mitigate spill damage by delaying approval of chemical dispersant use. It does not specify damages in its lawsuit. Exxon maintains the state had preauthorized the use of chemical dispersants in the oil spill contingency plan it had approved, thereby establishing a policy of using dispersants to respond to a spill. It accuses the ADEC of interfering with Coast Guard authorization to use dispersants and thus causing the subsequent delay that allowed the spill damage to get out of control.

That issue, along with determining the relative culpability of Hazelwood in the accident, is likely to be the most crucial to the outcome of any litigation stemming from the spill.

There has evolved a consensus that nothing could have contained the spill. Exxon itself has said that even if booms had been available, it is impossible to maintain enough spill containment equipment on hand to contain a spill of a speed and magnitude that could disgorge more than 150,000 bbl of crude into the water in less than an hour.

Further, placing containment booms around the crippled tanker in the early hours after the spill, as called for in the contingency plan, could have risked explosion and fire. Gas vapors had collected around the Exxon Valdez in the first two days after the spill. Placing containment booms around the vessel could have intensified the vapor cloud to a dangerous level.

Given the virtual impossibility—and added danger—of early containment, the legal issue after initial responsibility shifts largely to that of mitigation. Timing is the critical consideration. What if Exxon had obtained early approval for in situ burning and dispersant use?

Coast Guard Adm. Yost contended that dispersant approval would not have made a difference given the amount of dispersant Exxon had on hand—"a drop in the bucket," in his words. Could Exxon have obtained enough dispersant soon enough? Some early tests showed excellent results, but weathering of the crude later rendered the dispersant largely ineffective.

What about a burn? Some have suggested, in hindsight, that simply "torching" the spill—once there was no danger to personnel or the tanker with its remaining cargo of almost 1 million bbl of oil—would have been the most effective containment procedure. But what would be its environmental consequences? Simply a transient episode of polluting black smoke or an uncontrollable conflagration?

There remains an ancillary yet important issue that could dramatically affect the extent of damages awarded in Exxon Valdez lawsuits. The Department of Interior in August 1989 lost a court case involving pollution damages, one it did not appeal. A Washington, D.C., federal appeals court struck down an Interior regulation allowing companies guilty of pollution to pay the lesser of restoration or replacement costs. In other words, companies responsible for pollution under Interior regulations now must pay the full cost of restoring the environment to its original condition, not just the value of the damaged natural resources.

Full restoration, especially in an area such as Alaska, could be enormously expensive, especially if there is a stringent interpretation of the term.

That issue is at the heart of a separate lawsuit by the National Wildlife Federation, Natural Resources Defense Council, and Wildlife Federation of Alaska against Exxon and Alyeska. The environmental groups want Exxon and Alyeska to set up a $1 billion plus fund to mitigate environmental damages in Prince William Sound. In addition to seeking unspecified punitive damages, the environmental groups want an Alaskan Superior Court to oversee the fund and cleanup efforts to ensure that the sound and its wildlife are cleaned and restored.

Cleanup Status

• •

In midsummer 1989, Exxon and the Coast Guard recognized that the company would have to beef up its cleanup forces in order to meet the September 15 deadline (Table 1–1).

Table 1–1 Scope of the Exxon Valdez Spill Cleanup

Among the measures taken in its cleanup efforts, Exxon:

- Spent more than $2 billion, including more than $300 million in fishing claims.
- Treated 1,089 mi of shoreline in Prince William Sound and the Gulf of Alaska.
- Assembled a peak fleet of 1,436 vessels and aircraft.
- Deployed at peak more than 420,000 ft of boom, enough to encircle Manhattan Island three times.
- Collected more than 500,000 bags of oil debris that were shredded and sent to sites in the Lower 48 for disposal because of difficulty in obtaining incineration permits in Alaska.
- Developed systems for cleaning shores that could deliver more than 200 million GPD of water, enough to serve a city of 1 million people.
- Installed 1,000 industrial grade water heaters to supply almost 17 million GPD of warm water to oiled shorelines.
- Hired 20 archaeologists to help identify and protect artifacts threatened by the spill.
- Air freighted in clams and lobsters to feed sea otters at wildlife rescue centers at a cost of $100 per day per otter.
- Called about 30 former Exxon employees out of retirement to help in a variety of support tasks.

At its peak, the cleanup force totaled 1,436 vessels, 82 aircraft, and more than 11,000 persons. Deployment of boom in the sound was enough to encircle Manhattan Island three times at that time.

As damaging as the spill was to Alaska's fishing industry, the massive workforce assembled for spill cleanup drove Alaska's unemployment rate in August to a record low of 5.9%. In August 1988, Alaska's unemployment rate was 8.8%. During the six months of the cleanup effort, more than 15,000 people worked in some connection with the cleanup, about a tenth of which came from outside the state. In the areas hardest hit by the spill, dominated by the fishing industry, unemployment ran perhaps half of that the same time the year before.

The only other Alaska employment boom to compare with the spill cleanup was the construction of TAPS in the mid-1970s, a three-year effort.

As the deadline approached, the controversy over what would constitute "clean" heated up. Exxon backed off its claim of cleaning all the affected shoreline by September 15 after Alaska and environmentalist groups claimed a standard of restoration to a pristine state as the only viable standard for "clean."

Instead, Exxon's subsequent reports of cleanup progress referred to how many miles of shoreline had been "treated."

The controversy further heated after an Exxon official said that Exxon would return in the spring with a 20-person survey crew and would not commit at the time to a renewed cleanup effort.

That elicited another round of indignant howls. The White House and the Coast Guard were quick to guarantee Exxon's presence and continued cleanup efforts in the spring.

Admiral Yost, in an interview on ABC's "Good Morning America" show, said, "The job is not done, and it is going to have to be finished up next year. Exxon will be back in the spring, and they'll do what we tell them to do."

Exxon gave in to the pressure and submitted a winter program for operations related to the spill cleanup. The plan focused on maintaining 300 company and contractor employees in Alaska during the winter. About 20 of those were stationed along with 4 vessels and 7 aircraft in the spill area. Much of the winter effort was devoted to gathering data and maintaining a level of oil spill response capability.

The company's winter program also called for 26 major science projects to underpin additional surveys in the spring. The Coast Guard used data from those projects to determine what added cleanup would be needed in the spring.

Exxon claimed it accomplished what it set out to do: Remove enough of the gross oil contamination to make the area affected by the spill environmentally stable.

Exxon Chairman Rawl contended that most of the shorelines affected by the spill would be what most people would consider clean. Although one could still find some oil residue in a small portion of the affected shorelines, Rawl said, those areas were environmentally stable and posed no risk to fish or wildlife.

Admiral Yost contended that Exxon had underestimated the cleanup effort by about 20% and would have to finish the job in the spring. He predicted the sound would recover in a few years.

Exxon's claims of success brought a parade of officials and representatives of environmental groups to the area, accompanied by television cameras, to refute those claims. Perhaps the high point of this effort, or nadir, depending on one's point of view, was the visit to some rocky shore in Prince William Sound by Wilderness Society President Jay Hair and pop music star John Denver. This was about the same time Exxon and Alyeska were taking reporters on a tour of the demobilizing cleanup program in the sound.

Hair and Denver, accompanied by a platoon of videocams, walked along a shoreline they claimed had been treated by Exxon. They hoisted obviously oiled rocks and their subsequently oiled hands up to the cameras as "evidence" of Exxon's duplicity. Later that day, Hair and Denver held a press conference competing with a briefing by Exxon officials to company employees on the company's winter plans. When Denver was asked about surveying the sound's beaches and Exxon's claims, he said flatly, "They lied."

An Exxon official responded to that charge by pointing out that Denver and Hair essentially called the Coast Guard liars as well, since the agency had to walk and sign off every inch of beach Exxon claimed to have cleaned. What he did not point out was that there were several heavily oiled beaches in the sound that the EPA had asked Exxon to leave untouched as control beaches for testing accumulated oil effects. Knowingly or not, it could have been one such beach getting the pop star/media circus treatment.

Even as the first dusting of snow laced the Chugach Mountains along the sound in October, controversy continued to swirl over Exxon's winter plans and the effectiveness of its cleanup.

In addition to its wintertime contingent, Exxon kept another 250 contract personnel on call to supplement full-time staff for added contingency response capability. It also warehoused about 22,000 tons of equipment as part of that capability in Anchorage, plus extra response equipment in Valdez, Cordova, Whittier, Kodiak, Seward, and Homer and containerized cleanup supplies for hatcheries.

One cleanup effort continuing through the winter showed excellent results in June testing: bioremediation. The technique involves adding nutrients to oil to promote development of bacteria that feed on carbon. The bacteria break down hydrocarbon molecules, leaving carbon dioxide and water. In tests, Exxon had treated about 70 mi of shoreline in Prince William Sound. Bioremediation broke down surface oil and apparently removed some oil that soaked into the ground.

Meanwhile, Alaska announced plans to pursue its own wintertime cleanup efforts in a $21 million program, citing inadequacies in Exxon's winter program. Exxon denounced the state's program as vague and duplicative of its own effort. Otto Harrison, general manager of Exxon's Valdez operations, accused Alaska Gov. Cowper of playing politics, noting that the key difference in the two programs—operating an added six weeks—would be tantamount to playing Russian roulette with the safety of workers on the shorelines. That point was underscored shortly thereafter as a tug and an empty barge it was towing were lost in stormy seas along with three persons. The tug had just gone off charter after working in the cleanup effort.

As for cleanup efforts in 1990, Exxon expected to reach a recommendation on its plans for 1990 by the second half of March.

Prospects for the sound's recovery appeared good at midyear 1990. Early data from environmental studies of the sound suggested that damage to its ecosystem might not prove as severe as some had feared.

A midsummer 1989 study by the University of Alaska's Institute of Marine Service found that the sound's plankton bloom was advancing normally with large populations of animal and plant organisms.

The spill had affected at most 10% of the sound's linear shoreline and did not result in massive fish kills. Oil trapped beneath the surface was probably unlikely to return to the water column later, Exxon said.

A preliminary review by Exxon in January and February 1990 showed surface coverage of the oil throughout Prince William Sound and the Gulf of Alaska decreased as much as 75% in the same period. Total oil concentrations in the sediments decreased an average 75% in the same period. More importantly, biological communities, even on oiled shorelines, showed healthy, early repopulation and growth.

Exxon also studied the processes involved in the reentry of oil into the ecosystem and found that environmental risks were minimal and diminishing as weathering continued. Average hydrocarbon concentrations in the water column measured for more toxic components showed levels well below Alaskan standards and as much as 1,000 times lower than lethal levels for plants and animals.

As of April 20, 1990, 574 mi of shoreline had been surveyed. There was no evidence of surface oil on 372 mi, very light oiling on 130 mi, and a narrow to wide band of weathered crude on 72 mi. About 16 mi showed significant levels of subsurface oil.

Exxon continued cleanup in summer 1990 with small mobile squads of workers using least intrusive cleaning techniques. The teams used bioremediation for surface and subsurface treatment, manually picking up oiled debris and tarballs, removing tar mats, spot washing with hand-held units, and tilling and raking sediments. Eight teams were supported by 8 berthing vessels, 12 landing craft, 24 helicopters, 10 other aircraft, 20 workboats, 8 supply boats, and about 30 smaller boats and skiffs.

Furthermore, the weather of the sound itself contributed largely to the cleanup. Some of the beaches that were left as EPA control beaches were showing substantial signs of self-cleaning from action by waves and tides as the winter weather began hammering the rocky shorelines in late September.

One Exxon official confided that perhaps the best thing to do would be, simply, to "let Mother Nature have a couple of winters at it, and that would probably take care of most of the problem."

June Lindstedt-Siva, ARCO manager of environmental science, predicted in an interview with ARCO Environmental News Update, that most of the sound would recover in three years, give or take a year.

> "There have been some ominous predictions made in the press: that the sound will be dead for decades, and that the sound's fishing culture is over. I remember a similar prediction at the time of the Santa Barbara spill in 1969. A well-respected biologist said that the channel would be dead for 25 years. It recovered in two years.
>
> "At that time, spills had not been extensively studied, and no one knew what to expect. Since the Santa Barbara spill, many studies have been conducted, both in the laboratory and in the field.
>
> "We know a lot more about the impacts of oil than we did then and should be able to make a more informed estimate of recovery. Dire predictions of the demise of the fishing culture and decades to recovery are unsupported by any scientific evidence."

Exxon may well finish cleanup of all gross contamination in Prince William Sound in 1990, but the company is more cautious these days about making grandiose predictions. Exxon likely will be returning to the oiled areas for years to come to follow up with studies to determine the long-term effects of the spill on the sound's ecosystem.

The company will pay for the spill in many ways, probably for many years. It slashed its 1989 income with a $1.68 billion provision for the spill against earnings for the spill costs. The price tag in early 1990 for the cleanup alone had already topped $2 billion. Aside from claims and legal costs and the innumerable and intangible costs from legislation and regulation stemming directly and indirectly from the spill, there is the perhaps irreparable damage to Exxon's image.

The sound is likely to recover. Even Exxon and the Alyeska companies, after spending perhaps billions of dollars to clean up history's costliest nonlethal (at least to humans) disaster and fighting protracted legal battles through the 1990s, will manage to endure.

But what does the Exxon Valdez spill portend for the U.S. petroleum industry overall in the years to come?

Beyond the environmental damage, the staggering costs, and the self-inflicted pain of Exxon and Alyeska, on March 24, 1989, for the U.S. petroleum industry, the world changed forever.

2

THE MORE THINGS CHANGE . . .

Prelude to Change

. . . .

The year 1968 was one of the most tumultuous in modern U.S. history. Every day, the print and electronic media bombarded the American public with news of civil unrest, protest, riot, demonstration, war, and assassination. It was a time of sweeping change, a revolution under way in social mores and cultural values rocking the established order.

One of the currents of change was an emerging environmental consciousness, a nascent acknowledgement that man was fouling his nest.

In a May 10, 1968, special essay titled "An Age of Effluence," *Time* magazine took note that "scholars of the biosphere are seriously concerned that human pollution may trigger some ecological disaster."

The irony in the essay's title was evident enough. The 1950s were already tagged the "Age of Affluence" and viewed even in 1968 with derision as a period of almost soporific complacency, conformity, and rampant consumer materialism. An abiding article of faith in the 1950s was trust in one of the key institutions: big business. With the socio-political revolution of the 1960s and its focus on anti-materialism, it was just as much an article of faith that such guiding institutions were not to be trusted.

The publication of Rachel Carson's *Silent Spring* in 1962 is largely viewed as the birth of the modern environmental movement. Her analysis of the detection of the widely used pesticide DDT in the food chain and the chemical's harmful effects on wildlife and potentially on man was probably the first widely read indictment of an institution sewing the seeds of environmental destruction.

Still, ecological concerns were largely seen as the stock in trade of a fringe element.

Even the disastrous Torrey Canyon tanker oil spill that spewed 715,000 bbl of Kuwaiti crude oil along the coasts of England and France had failed to ignite the kind of galvanizing passion needed to forge a new movement in the crucible of public opinion. Perhaps the consensus of institutional mistrust had not fully formed yet. And after all, the Torrey Canyon spill happened "over there." Such things were pretty unlikely in the United States.

The war, civil rights, and other issues held center stage in the public's consciousness in 1968. But a string of largely unrelated events and scientific studies that year kept concern for ecology nibbling at the edge of citizen awareness.

There were reports that the presence of DDT in the food chain was probably responsible for alarming declines in the populations of peregrine falcons, black-crowned night herons, and other bird species in New England.

Recognizing that motor vehicles account for 60% of all air pollution, Congress empowered John W. Gardner, Secretary of Health, Education, and Welfare to limit auto pollutants. Proposals before Congress called for eliminating 77% of the emissions of hydrocarbons and 68% of carbon monoxide by 1970. The Department of Interior's Bureau of Mines started testing a new exhaust manifold reactor designed to reduce tailpipe emissions.

A nine-day garbage strike in summer afflicted New Yorkers with mountains of trash accumulated on sidewalks, reminding the public of the enormous volume of refuse it generates and the declining availability of disposal options.

The Pentagon released a study of the use and effects of herbicides as defoliants in Vietnam. Chemical defoliation had stirred up controversy among U.S. ecologists, some of whom maintained the practice threatened several rare animal species in Indochina. Another concern was defoliation's resultant effect of laterization, in which the soil in defoliated areas becomes exposed to erosion and sunlight, growing rock hard and eventually useless.

The U.S. Food and Drug Administration began efforts to ban carbon tetrachloride after studies showed the widely used cleaning solvent, when absorbed through the skin or inhaled, caused a variety of side effects and illnesses, sometimes death.

There was no lack of Doomsday predictions in 1968 owing to obscure or esoteric scientific phenomena. A study by the Environmental Science Services Administration and the U.S. Geodetic Survey reported a decline of 15% in the earth's magnetic field in the past 300 years. They predicted the earth's magnetic field could disappear by 3991. The result: an intense solar bombardment that could cause icecaps to melt, coastal cities to be flooded with the rising sea level, lush areas to grow barren, and other nightmarish scenarios.

Studies coming out on a regular basis illustrated mounting concerns in a wide variety of environmental issues: worsening smog in New York, with the memory still fresh of 1963, when a thermal inversion trapped smog that was blamed for killing 400 New Yorkers; rising death and illness rates attributed to the presence of lead in the air from automobile exhausts; Lake Erie pronounced more or less officially dead because of generations of effluent dumping into the lake; prognostications of a new Ice Age, foretold by several consecutive years of progressively colder temperatures; poison in the nation's water supply from runoff of fertilizer and pesticides from agricultural operations; and the first recommendations that consumer goods manufacturers be required to introduce rapidly degradable packaging. These were among the issues that led some to recommend that the Secretary of the Interior instead be renamed Secretary of the Environment.

All the elements were in place for widespread change in public consciousness—transforming a motley assortment of fringe issues into a grassroots movement to effect a fundamental change in society. All that was needed was a catalyst.

It came on January 28, 1969.

The Catalyst, 1969

. . . .

Drilling for and producing oil in the Santa Barbara Channel in the 1960s was hardly a cakewalk.

Some local opposition to oil development, particularly in the city of Santa Barbara, had been a fact of life in the region since the turn of the century. The city had long been noted for fierce opposition to any large industrial development, including stopping a major overhead freeway proposal. But there also was a healthy contingent of oil development supporters, notably in nearby coastal communities like Ventura and Hueneme that benefited from the jobs and revenues channel oil development provided.

The state was pretty clear how it felt about offshore oil development. It was embroiled in a dispute with the U.S. Navy and Air Force over allowing oil and gas leasing in areas of the channel that the military branches wanted reserved as danger zones for missile test firing.

A State Lands Commission report on the subject at the time noted, "Approximately 118,000 acres of state-owned tide and submerged lands in presently established danger zones have been foreclosed from state management and development of potential natural resources. The state thus already has been forced into economic losses from restricted offshore development."

Despite military opposition, the commission in January 1969 was making plans to offer a parcel off Point Conception near several other oil discoveries. The state also was moving ahead with plans to offer oil and gas leases in San Pablo Bay, although several cities and counties in the San Francisco Bay area objected, contending oil operations would prove a hazard to recreational boating.

The California Water Resources Control Board was to rule by early February on a dispute between the oil industry and water conservationists concerning the water quality of Long Beach harbor. More than 250 wells in giant Wilmington field, which had been discharging brine wastewater into the channel, would have to be shut in if proposed tighter water pollution controls were to be enforced in order to render the channel capable of sustaining fish life.

In the dispute, the new chairman of the water board appeared to come down on the side of industry, noting that of all entities the oil industry had spent the most money to solve water pollution problems and that he saw an increasingly unfair tendency to brand oil companies as villains in pollution issues.

Meanwhile, the oil industry faced political threats in two key issues on Capitol Hill in addition to the perennial one of higher taxes: refinery safety and water pollution, notably oil spill legislation.

These developments played out against the backdrop of heightened concerns over world crude oil supply and demand. A Georgetown University study warned about the repercussions of a British withdrawal from the Persian Gulf. The study said that a cutoff of Middle East supplies could have serious strategic, political, and economic consequences around the world. It recommended that consuming countries develop large crude oil stocks and new sources of supply.

Concerns over U.S. energy supplies also were mounting. In 1968, for the first time, U.S. demand for oil surpassed domestic supply.

Critical to the future crude supply of the United States was the giant discovery at Prudhoe Bay on Alaska's North Slope. As a dozen operators were busy delineating the field's limits beneath the frozen tundra in 1968, analysts were already estimating the find as holding perhaps 5–10 billion bbl of oil, making it the biggest oil discovery ever in North America.

Rising import levels also contributed to the federal government's push to lease the Outer Continental Shelf. A February 1968 lease sale in Los Angeles netted the government more than $603 million in cash bonus bids at the time. Oil platforms already were proliferating in the Santa Barbara Channel.

Industry and federal officials had taken pains to assure Santa Barbara area residents that no leasing or drilling would occur without strict safeguards to protect the environment.

The Blowout Heard 'Round the World

Union Oil Co. of California had installed two platforms in about 180 ft of water in federal waters about 6 mi south of the city of Santa Barbara.

The platforms were installed on federal tract 402, for which Union, Mobil Oil Corp., Gulf Oil Corp., and Texaco Inc. paid more than $61 million in a federal lease sale the year before. Union was operator of the leases for the group.

In the late morning of January 28, 1969, drilling contractor Peter Bawden Drilling Co. drilled the fifth development well, 402–A–21, from Platform A on Tract 402 to about 3,500 ft. The contractor set about 514 ft of 13⅜-in. casing 254 ft below the mudline. The crew was coming out of the hole and preparing to run an electric log when the well began to kick. A geyser of oil and mud suddenly rushed out of the well, spewing 100 ft into the air over the rig derrick.

The rate of oil flow prevented success in stabbing the kelly, so the crew dropped the drill pipe into the hole in order to close the blind rams in the blowout preventer.

Had it not been for the quirky geology of the Santa Barbara Channel, this would have solved the problem, and industry history would have been decidedly different. However, a history of strong seismic activity left the subsurface strata in the channel laced with a network of faults and folds. Oil and gas seeps dotted the channel floor.

Minutes after the well was shut in, oil and gas began to percolate to the water surface, apparently from a subsurface fault near the wellbore. Boils of oil, gas, and silt appeared in five locations as far as 800 ft from the platform. The wild well was disgorging oil at the rate of about 500 B/D into the channel.

Gas escaping from the well at first prevented efforts to reestablish circulation in the well to kill the leak. Union hired a nearby drilling barge to drill a relief well about 1,000 ft southeast of the platform. It also began efforts to contain the oil spill, spraying Corexit onto the slick, deploying plastic booms, and dumping straw around the edge of the slick.

Union brought in the famed blowout specialist Red Adair to kill the well. Gas fumes and continuing storms hampered Adair's efforts to kill the well. After 11 days, he was finally successful. But for three months thereafter, oil continued to seep in thin stringers from seafloor fissures near the platform. By that time, more than 50,000 bbl of crude had fouled the channel.

Initial oiling of the coast was light. But shifting winds drove the slick into a 12-mi stretch of the Santa Barbara County coast. Within a week, the slick had spread to cover an area of 800 sq mi. Oil was coming ashore as far south as Santa Monica and Malibu.

Television footage of the environmental devastation stunned the nation. Newspapers covered details of the cleanup every day, quoting Santa Barbarans in full fury. *Life* magazine's photo essay depicting heartbreaking scenes of oil-soaked seal pups dying in agony next to their frantic mothers bore mute testimony to the grim toll on wildlife.

Immediately, the political fire ignited by the A–21 blowout was whipped into a maelstrom. Environmentalist groups and Santa Barbara County officials demanded President Nixon halt all drilling in the channel. A state senate resolution supported by Gov. Ronald Reagan called for on-site inspection of all offshore oil operations in federal waters. Interior Sec. Hickel banned new drilling in the channel pending a federal investigation and flew to Santa Barbara to inspect the site. A temporary ban on current drilling lasted only a few hours until companies could comply with procedural changes requested by the U.S. Geological Survey. The State Water Resources Control Board was asked to seek suspension of all drilling off the coast. The State Lands Commission began reviewing drilling rules in state waters.

The repercussions were just as bad in Washington. President Nixon and Sec. Hickel called for new regulations governing drilling on the Outer Continental Shelf, and Hickel said no more steps toward leasing would occur until drilling practices were reviewed.

The industry tended to regard the blowout as something of a freak accident. Union Oil President Fred L. Hartley told a Senate committee shortly after the blowout that it probably would not have happened had the well been cased to 2,000 ft. The well had been open from 500 ft at surface casing to the producing zone at 3,500 ft. No one, Hartley said, could have guessed beforehand that Interior-approved drilling practices were insufficient in such a situation.

The *Oil & Gas Journal's* Washington Editor at the time, Gene Kinney, saw the handwriting on the wall in the magazine's first issue after the spill:

"The huge oil spill in the Santa Barbara Channel will be reverberating through the industry long after the offending well has been tamed. It will be the catalyst which shapes events for years to come."

Even as cleanup crews on the California beaches swelled to 1,500, and initial fears that the channel would be dead for years were beginning to abate, there were tremors of a new militancy being felt in California, and its epicenter was Santa Barbara.

The long simmering resentment and distrust towards the oil industry in Santa Barbara had boiled over into an explosive anger that soon was channeled into a call for action.

More than half of the estimated 1,100 birds oiled in the spill had died, many agonizingly in the hands of longtime residents acting as volunteers in the cleanup. Hundreds of boats were damaged and commercial fishing operations disrupted. Lawsuits totaling more than $3.5 billion were filed against Union in the first month after the beginning of the blowout.

The sense of righteous anger coupled easily with the newly found social activism in Santa Barbara. Long a tourist haven, historic site, and sort of bedroom community for Hollywood types seeking respite from the rigors of their glamorous calling, Santa Barbara bore a natural enmity for any industrial presence. Some of the city's residents banded together to form a grassroots organization to "Get Oil Out" once and for all. And GOO was to be the touchstone of a new force in the United States, one that would change the face of American politics forever.

GOO was formed as a result of a public rally at Stearns Wharf, where the local populace gathered to denounce Big Oil, declaring the following Monday "Black Monday," during which motorists were urged against buying gasoline or motor oil. GOO followed up with letters of protest to Washington representatives and officials in Sacramento. They also staged a public burning of oil company credit cards and boycotted gasoline stations connected with companies participating in drilling in the Santa Barbara Channel. Other protest actions included forming car pools and walking, riding, or taking the bus.

Subsequent U.S. Senate hearings on the spill were jammed with press and televised. Santa Barbara residents were called before a subcommittee on air and water pollution to testify on the spill and its aftermath as the television cameras rolled. Sec. Hickel, who previously had been deemed a pawn of oil companies at his confirmation hearings, was accused by one Santa Barbaran of cynicism and hypocrisy in initiating and then quickly ending the ban on current drilling with the implementation of new safety procedures. The hearing later took on overtones of a TV drama as the committee chairman Sen. Ed Muskie, tall and Lincolnesque, verbally jousted with the pugnacious, stocky Union Oil president Fred Hartley.

After the subcommittee hearing, the nation's newspapers carried front page stories that quoted Hartley as telling the committee members, "I'm amazed at the publicity for the loss of a few birds."

The outrage was instantaneous. For sheer arrogance and corporate insensitivity, this had no peer, hearkening back to the rapacious Robber Baron mentality, pundits said.

The irony was that Hartley never said it. A paraphrased summary of Hartley's testimony, perhaps with some sarcastic embellishment, was given by a *Christian Science Monitor* reporter to a *New York Times* reporter who had not been there for Hartley's testimony. The paraphrase ended up as a direct quote in the *Times,* in turn subsequently lifted by the *Wall Street Journal* and repeated hundredfold across the nation.

An enraged Hartley demanded and got retractions, but they were ignored. The damage had been done. And the flames of militant environmentalism were fanned still higher.

There were other developments throughout 1969 to keep environmental issues at the forefront of public consciousness. A graphic example of environmental degradation became the stuff of editorial cartoons and "Tonight Show" monologues: The Cuyahoga River, which splits Cleveland, Ohio, and had suffered years of municipal sewage runoff and industrial dumping, caught fire several times.

Washington Democratic Sen. Henry "Scoop" Jackson introduced legislation that was to become the National Environmental Policy Act of 1969. The measure provided that environmental as well as economic and technical considerations govern the decision-making process of federal agencies. It introduced a new phrase into the American businessman's lexicon—one that would carry increasingly heavy freight through the years: environmental impact statement. Now any legislation or administrative action to be carried out by a federal agency would require a study of the various potential adverse or beneficial effects on the environment. Those findings in turn would be subject to public hearings.

Time magazine, always the barometer of our society's trends, in 1969 added a new department: The Environment. Reporting on the growth of a new type of student activist that had begun picketing polluters and campaigning for antipollution laws, *Time* spoke of ecology as an emerging "demi-religion."

Longtime social activist and folk singer Pete Seeger took a broadly publicized antipollution protest voyage on a schooner down the Hudson River in the summer of 1969. That was one of the early social protest "photo opportunities" by a celebrity who was to have a direct descendant 20 years later in John Denver hunched over oily Alaskan rocks and showing his soiled hands to the camera like a toddler proudly showing off his proficiency at making mudpies. The environmental movement would have no better tool than the electronic media.

Another catastrophic oil spill, the two-year-old Torrey Canyon tanker spill, would make the headlines again in 1969, at the expense of beleaguered Union Oil. Union and the Liberian owner of the tanker in November paid $7.2 million to settle claims by Britain and France related to damage and economic losses caused by the spill. They had filed claims totaling $15 million.

Nor would the Platform A blowout be 1969's only oil spill. Later that year, the French tanker Gironde rammed an Israeli vessel, spilling 7,000 bbl of crude off the coast of Brittany. And the British tanker Hamilton Trader collided with another vessel, spilling oil off Wales.

In Santa Barbara, the luckless Union was involved in other incidents that further stained its image. Even as Sen. Muskie's subcommittee heard testimony in Santa Barbara about a month after the initial blowout, news of reputed further leakage—apparently the result of downhole work in the relief well—reached the subcommittee.

The night after the hearing, lightning struck Platform Hilda, operated by Standard Oil Co. of California, in state waters about 4 mi from Platform A. And the following day, the Santa Clara River jumped its banks and carried three Union storage tanks into the channel.

Beach cleanup work continued through much of the year as the channel's natural seeps and leakage from new seafloor fissures perpetuated the belief among Santa Barbarans that spills were continuing routinely.

Just before Christmas in 1969, a hapless Union was thrust into the spotlight again when a rupture in the pipeline linking Platform A and the shore caused some additional minor leakage.

It was beginning to look as if nothing would go right for the industry. The string of other spills, together with a rash of blowouts and other industry accidents left the public with an indelible image of the petroleum industry as a blundering giant always ready to put profit before environmental and safety concerns. And the fires of eco-evangelism grew white hot.

On the first anniversary of the oil spill, January 28, 1970, GOO coordinated a cross section of groups into an Environmental Rights Day at Santa Barbara. Only days before, President Nixon had asked for $10 billion for pollution control, terming it the nation's No. 1 priority.

The environmental movement sparked by the spill swept across the country, peaking on April 22, Earth Day, designated as a day devoted to national concerns for the environment.

Outside the war in Vietnam, nothing else had so galvanized social unrest into a cohesive movement as had the environmental movement. Even the war did not cut across demographic groups as environmentalism did. There were deep divisions over the rightness or wrongness of the war, but *everyone* supported care and respect for the environment. And here was an easy villain: corporate greed in the guise of the oil industry.

In the years to follow, environmentalism was to transform the fabric of American life. Now there was a potent new political force abroad in the land, creating a bewildering array of agencies, committees, regulations, review, and a welter of bureaucratic layers. Only the wild gyrations of oil prices in the 1970s and 1980s had greater effect on the petroleum industry than the staggering costs of environmental mitigation ultimately traceable to the Santa Barbara spill (Fig. 2–1).

Had it not been for skyrocketing oil prices after the Arab oil embargo in 1973, it is likely that development of supergiant Prudhoe Bay oil field and

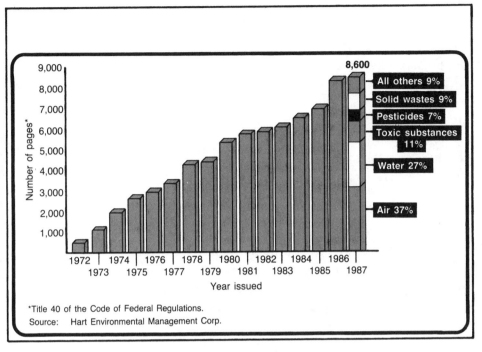

Figure 2–1 How federal environmental rules mushroomed.

construction of the TAPS would have been killed by environmental activism. It was only a few votes in Congress, spurred by an energy crisis and gasoline lines, that made the difference even then.

Environmentalism permeated every fiber of the American political fabric in the 1970s, spawning cabinet level agencies and whole new industries devoted to mitigation, cleanup, and study. A powerful new political lobby was flexing its muscle in Washington, D.C., and especially in California. Environmentalism soon became big business and a potent constituency.

With the Reagan years, the growth in environmentalism slowed. Part of that could be laid at President Reagan's doorstep, with his vow to press for deregulation and getting the government off the backs of American citizens and businesses. The regulatory rollback on environmental issues was aided in the beginning by the devastating effects of oil prices that had turned the world economy topsy turvy and were thought headed for $50, perhaps $100, per barrel this century. Another contributing factor was that the oil industry's environmental record since the Santa Barbara oil spill was impeccable: no lasting damage from the Santa Barbara spill itself, no further significant effects from offshore drilling and production operations, none of the environmentalists' predicted severe damage from drilling, producing, and transporting oil in Alaska's pristine wilderness.

There were even new lease sales off California—record-setting affairs—and monster discoveries, including the biggest ever off the United States at Point Arguello field. Despite the heated controversy that persisted, even the state allowed (after extravagant mitigation measures were taken) oil companies to drill in state waters again.

With the collapse in oil prices in 1986 came a corresponding fall in gasoline prices. The resulting economic dislocations in the oil patch states elicited sympathy from the rest of the nation, given a booming economy elsewhere and the fact that motorists found gasoline the best bargain around. Not in many years had the industry such a favorable public image.

The underlying message from consumers and voters seemed to be: Keep your act relatively clean and the price of gasoline down, and you can drill pretty much where you want, with some notable *Not-in-My-Backyard* exceptions.

The Reagan years also followed the Carter "malaise" presidency. Citizens were thoroughly weary of all the carping about America, the attacks from within as well as from abroad. A few missiles lobbed at a tin-pot dictator, a quick little invasion to stop the Red Menace close to home, America's space program aloft again—all helped revive nationalist sentiments. John Q. Public was starting to develop a little more confidence about America, even a little complacency about the "can-do" spirit of American business, technology, and political and social institutions.

With the late 1980s, notes of doubt about American institutions began to creep in again: Iran-Contra, the Challenger space shuttle disaster, spiraling trade and budget deficits, soaring drug and crime crises, the once-burgeoning fundamental Christian movement crippled by scandal, growing ineffectiveness of foreign policy—all were hallmarks of the Reagan regime's waning years.

The country's newly rediscovered malaise was hardly helped by a presidential campaign that offered a pair of nominees from the two major parties generally perceived as lackluster. Vice-President George Bush sought to score some early points on his Democratic opponent, Massachusetts Gov. Michael Dukakis, with a photo opportunity to serve as a backdrop for charges that the governor's disinterest or bungling had prevented a cleanup of the polluted Boston harbor.

Bush, eager to establish an image for himself other than as a perennial jack of all trades to various presidents, had injected a note in American politics that had scarcely been heard since the 1970s: He wanted to be known as the *environmental* president. It was a linchpin for 1988 to become, *The Year the Earth Spoke Back.*

The Year the Earth Spoke Back

. . . .

Americans had ushered in 1988 with an upbeat mood. Although hardly free of concern, they seemed to exude the continuing mood of confidence and optimism about the future they had held for much of the Reagan years.

A *U.S. News & World Report* poll surveyed Americans to gauge their chief concerns for 1988 in the magazine's "1988 Outlook" issue. At the top of the list were worries about the spread of AIDS and the crack cocaine epidemic. Environmental concerns did not even make the list. It was not as if environmental issues no longer mattered; they simply were crowded out by other concerns. Although the oil industry was fighting pitched battles on several fronts, such as the campaigns to gain access to ANWR's Coastal Plain and proceed with development plans and further leasing off California, there was no single environmental issue that preoccupied the American consciousness.

Perhaps Americans had become somewhat blasé about the steady parade of studies and experiments that warned of dire threats to health and environment from all manner of things. It had become something of a cliche that "Everything causes cancer!" It may be that the complacency of the Reagan years also included the "Boy Who Cried Wolf" syndrome. That was furthered by an EPA study that suggested it was time to reconsider what the relative risks were from exposure to certain chemicals. In reviewing the 65,000 chemicals that EPA routinely monitors, the agency found that almost all were proving less dangerous than previously believed. Cancer risks for some substances were found to be a fraction of what was thought before.

Energy was scarcely anything for the average consumer to worry about in early 1988, either. In the waning weeks of 1987, overproduction by OPEC nations caused oil prices to slide by $2.50/bbl. At the same time, 15 states scrambled to take advantage of a new federal law allowing for higher speed limits on roads other than interstate highways for the first 20 states to qualify.

It followed a similar move in 1987 authorizing states to raise their speed limit on rural interstates to 65 MPH from the 55 MPH limit that the energy crisis of the 1970s had ushered in. Thirty-eight states took advantage of the earlier provision.

It is likely that the rollback of the widely disliked 55 mph speed limit and the resurgence of muscle cars in the late 1980s had as much to do with the economic buoyancy and renewed self-confidence seen in the American people during Reagan's second term as it did with the halving of gasoline prices.

A new public health scare emerged early in 1988: radon gas. A National Academy of Sciences (NAS) report projected frightening cancer risks postulated for persons living in homes where the odorless, colorless radioactive gas had collected in excessive levels. The study and subsequent press reports gave rise to a cottage industry of radon-testing kits and services for those concerned that they might be afflicted by an otherwise undetectable environmental threat.

Some might say that the environmental movement did not disappear in the 1980s, it merely regrouped in California during the 1980s. Evidence of the movement's continued strength in that state was clear early in 1988, when Proposition 65 labels began appearing everywhere. The Prop 65 initiative, which was approved in 1986 and took effect in 1988, forbids the discharge of substances thought to cause cancer or reproductive damage into or near drinking water supplies. Although its interpretation remains much in dispute, it provides a chilling precedent for industry with a "bounty hunter" provision that allows anyone to file a claim of discharging carcinogenic or teratogenic substances against anyone else. The informant is encouraged with a reward of a percentage of fines collected.

Concerns mounted for several years over the depletion of the ozone layer in the stratosphere. A massive study completed in 1988 showed that levels of ozone had fallen by an average 2.3% over North America. The study, which had the participation of more than 100 scientists around the world, estimated that for every 1% decrease in the ozone layer—which shields the earth from harmful ultraviolet rays from the sun—the incidence of skin cancer is likely to increase 5–7%, or about 500,000 additional cases each year. The new study's results hinted that the problem was far worse than had previously been believed. The culprit: release of halogenated chlorofluorocarbons (CFC) into the atmosphere. The U.S. Senate in late March 1988 unanimously approved an international treaty to slash CFC use by half in the next 10 years.

The summer of 1988 saw the worst drought in 50 years in North America. Record heat waves and dry spells afflicted much of the United States, ravaging crops and livestock in much of the West, South, and Midwest. Billions of dollars in crops were lost, fires raged throughout the West, and millions of acres of topsoil were blown away by relentless winds. Water shortages turned up in western states, compounded by outmoded utility monopoly regulations. Some climatologists claimed the brutal drought and heat wave was a harbinger of things to come. They predicted a gradual warming in global temperatures because of the greenhouse effect, in which the collection of certain gases in the atmosphere traps solar heat there, causing an inevitable warming of the earth's climate.

The debate reached a turning point with the testimony of James Hansen, an atmospheric scientist with the National Aeronautics and Space Administration, before the U.S. Senate. Basing his conjectures on meteorological data showing the highest global temperatures in 130 years of record keeping, Hansen predicted further increases in world temperatures, perhaps by 3–9 degrees by the middle of the next century. These predictions conjured up an apocalyptic vision of hotter, drier summers, more severe droughts in now-temperate regions, gradual melting of polar ice caps and glaciers, and eventual rising sea levels causing flooding of coastal regions. "The greenhouse effect," he said flatly, "is here." The culprit: mainly burning of fossil fuels and widespread deforestation. Although the theory had been put forth often in the past, the bleak pictures from the parched farmlands of America gave tangible evidence to many that the crisis was at hand.

If the Dust Bowl images were not enough, the drought sparked a massive conflagration in one of America's most cherished landmarks: Yellowstone National Park. More than 100,000 ac were charred and the Department of Interior had a fresh controversy in how it handled the park fires, opting at first for a "Let it burn" approach in the idea that fire is part of the natural renewal process in forests. Public outcries spurred a reversal, however.

Environmental groups used the dramatic fires at Yellowstone to draw attention to Interior's policy on multiple uses for national parks. They decried the growing real estate development, oil and mineral leasing, dams, and logging along the boundaries of national parks.

Renewed concern—and increasing enmity between the United States and Canada—over acid rain prompted several new bills in Congress to curb the problem. Thought to be the product of sulfur dioxide and nitrogen dioxide emissions from industry and power plants in the U.S. Northeast and Midwest that mixed with moisture in the upper atmosphere to form a highly acidic precipitation, the phenomenon has been blamed for killing forests and lakes in the northeastern United States and Canada. Tough new standards to curb acid rain-related emissions were promised by both major candidates for President.

Midsummer 1988 also bore witness to the consequences of generations of dumping trash in the ocean. For several weeks in July, beaches along the New York and New Jersey coastline were closed as beachgoers were driven away by raw sewage, grease balls, and hospital waste—including syringes and vials of AIDS-tainted blood—that came in with the waves washing onto the beaches. The scene was to repeated often during the summer.

The beach closures sparked another environmental crisis cover story for *Time:* "Our Filthy Seas," tallying a new list of environmental degradation in the oceans: Agricultural runoff of fertilizers, topsoil, and pesticides carried by rivers into estuaries; wastewater from factories and sewage treatment plants dumped offshore; oxygen-depleting algae blooms spurred by excess nitrogen and phosphorus in the water; offshore littering of plastic materials that kill millions of birds, fish, and marine mammals every year; and oil spills; all adding up to another ecological horror story.

Worries about U.S. food supplies were revived by the Alar and Chilean grape scares. The resulting hysteria almost crippled the American apple industry and a resurgent South American nation's economy. Celebrity testimonials helped revive grassroots consumerism and opposition to pesticide use in American produce.

A report from the Center for Plant Conservation predicted that as many as 680 plant species could disappear in the United States alone in the next decade because of encroaching human populations.

The oil industry was beleaguered through the year with a series of accidents that caused massive damage and loss of many lives.

The year was scarcely two days old when an environmental crisis stunned the nation. A 40-year-old, rebuilt Ashland Oil Co. oil tank on the banks of the Monongahela River at Floreffe, Pennsylvania, suddenly ruptured, spewing out about 90,500 bbl of diesel fuel. Although a dike around the tank contained about 59,500 bbl and subsequent cleanup efforts recovered another 9,500 bbl more, the remaining 21,500 bbl escaped into the Monongahela. The oil slick subsequently entered the Ohio River, moving 20–25 mi per day and threatening the drinking water and ecosystems of the Ohio River Valley. At least 70 municipalities in West Virginia and Ohio stopped drawing water from the river and imposed conservation measures. The spill dissipated completely within a few weeks. Ashland cleaned up the resulting damage and prepared for a wave of lawsuits as well as dealing with questions over how it happened.

In spring 1988, a fire at Chevron Corp.'s Richmond, California, refinery injured nine workers, damaged the plant, and cut West Coast product supplies for a short time. The incident later resulted in charges and penalties totaling $877,000 by the Occupational Safety and Health Administration filed against Chevron.

The oil industry was at the center of one of the most spectacular disasters of 1988: the July 14 explosion and fire that ripped apart Occidental Petroleum Corp.'s Piper Alpha platform in the U.K. North Sea, taking with it the lives of 167 men. It was the worst accident in the 25-year history of North Sea oil operations and triggered a thorough investigation and revamp of industry safety practices in the North Sea. Losses were expected to run into several billion dollars. The incident also contributed to a tightening of world oil prices.

That same month, an explosion and fire in the 90,000-B/D fluid catalytic cracker at Shell Oil Co.'s Norco, Louisiana, refinery killed seven employees, destroyed the unit, and caused widespread damage throughout the town. It also contributed to a tightening of gasoline supplies, driving up prices.

The oil industry's image on safety was already suffering in the months leading up to the Exxon Valdez oil spill.

In December 1988, a tug rammed a barge off Washington, spilling almost 6,000 bbl of oil into the state's scenic Gray's Harbor, oiling parts of Olympic National Park and Canada's Pacific Rim National Park.

Only weeks before that, a 60,000 bbl spill of fuel from a supply vessel contaminated the pristine shores of Antarctica.

In an ominous harbinger, just three weeks before the Alaskan spill, the Exxon Houston tanker broke from its mooring at an Oahu, Hawaii, terminal, and floated aground on a reef. The accident caused a spill of 600 bbl of crude from an offloading hose and 200 bbl of bunker fuel from a damaged fuel tank.

The Exxon Valdez spill was to be soon followed by a string of other oil transportation accidents off the United States in rapid succession.

In one stunning three-day period in April, three oil spills hit U.S. coasts, off Rhode Island, Florida, and Texas. The woeful industry marine transport track record persisted through the year. In another one-week stretch in September 1989, an empty Exxon tanker lost power and drifted for an hour in the pristine Strait of San Juan de Fuca off Washington; 40,000 gal of gasoline spilled in New York's East River when a tank barge ran aground; and an unknown party spilled as much as 2,000 gal of fuel oil in a "hit and run" spill that closed 30 mi of the Hudson River.

In another year, this string of coincidental incidents might not have gotten much notice. But they came on the heels of the "Year the Earth Spoke Back," a year of doomsday warnings of wholesale environmental destruction given weight by searing images on television and in the print media of cattle dying on the parched rangelands, devastating forest fires, garbage-strewn beaches, dying forests and lakes.

History and science may well judge 1988's ecological catastrophe-mongering as nothing more than a collection of unfortunate coincidences tied into almost whimsical metereological change and used by opportunists to feed a growing hysteria. Certainly some of the petroleum industry's rash of accidents can be laid at the door of an industry stretched too thin by the ravages of the oil price collapse. And much of the interest in environmental concerns could stem from a recognition that the White House soon would get an occupant more receptive to those concerns.

In any event, the stage was set. These disparate, sometimes tenuous concerns were blending together to form an elemental stew of impending change. Just as in 1969, all that was needed was a catalyst. It could have been something else, but fate dictated it would be the Exxon Valdez.

Had it been another company, the damage might have been ameliorated. Exxon was the ultimate icon to Americans of Big Oil: huge, bureaucratic, impersonal. Another company might have handled the public relations backlash better. It seemed to matter little at the time that probably no other company in the United States could have mustered the enormous resources to combat the spill that Exxon did. What mattered now was image, not substance.

And the environmental movement, not dormant, but certainly low profile on a national scale in the Reagan years, seized on that image with a vengeance. What environmental groups and their political surrogates are attempting now and will press throughout the remainder of the century is no less than an ambitious agenda to change the American consciousness.

There were already elements in the American collective mind that produced fertile soil for the seeds of such change—in part a legacy of the environmentalist and consumerist rebellions of the 1970s.

The result is that there is a new consciousness abroad in the land, and the petroleum industry must understand what it will come to mean in the 1990s, if it is to survive.

3

THE NEW CONSCIOUSNESS

In October 1989, the Doomsday Clock changed.

The clock's hands remained at six minutes before midnight—representing the world's proximity to the final midnight of nuclear destruction—despite the progress that year in diminishing tensions between the United States and the USSR.

That is because the Doomsday Clock began factoring in the global environmental threat in its simple symbol of Armageddon.

The Doomsday Clock, the logo on the cover of the Bulletin of Atomic Scientists since 1947, now appears superimposed on a globe.

With the threat of nuclear war easing under U.S.–Soviet rapprochement, the bulletin—begun as a newsletter among Manhattan Project scientists—has changed its view of what constitutes the main threats to international security.

Citing acid rain and aging nuclear reactors as well as the Bhopal and Chernobyl disasters, the bulletin's editors and its consulting international board of scientists made the change in focus to recognize that environmental issues are emerging as the most conspicuous threat to world survival.

This is but one of countless such changes reflecting the fundamental shift in the public consciousness on environmental concerns.

Earth Day 1990 will have come and gone as this book goes to press, but it is probably safe to say that it will have the same effect its counterpart of 20 years ago did. Earth Day 1970 involved 25 million Americans and led directly to the Clean Air and Clear Water acts, the EPA, and the defeat of a handful of congressmen targeted because they accepted campaign contributions from purported major polluters. Those and scores of lesser ripples from the Major Media Event of April 22, 1970, served to permanently enshrine the environmental ethos into the stew of cultural norms and mores that define America.

Think for a moment about how far American media has progressed since 1970 (some might say regressed) in saturating the public's consciousness with information when something becomes a "Big Issue."

Think how often that information has been served up as hype, generously larded with spectacle for the hungry television cameras.

Finally, think about how scientific or economic issues are presented in this context: They are either inscrutable and therefore boring, or, if placed in the context of person or pocketbook, are deemed a Great Threat or a Great Boon. Americans did not buy into space exploration until Sputnik and the Red Threat to conquer the moon put it into a concrete context, with specific, emotion-rousing images. Correspondingly, falling U.S. oil production and soaring oil imports were issues that gathered dust in the collective consciousness until Americans had to sit in line to buy gasoline for almost $2/gal. The issue became real when it became tangible in a personal way, assailing Americans' credo of Convenience With Cheap Commodities and, more importantly, threatening to hinder the daily worship of the Great God Automobile.

That is how the Year the Earth Spoke Back and the Exxon Valdez, like the Santa Barbara Channel oil spill, with their powerful images of environmental devastation, served to concretize the New Environmental Threat as something very real and very urgent to the average American. It was left to Earth Day 1990, ironically coming close to the first birthday of the Exxon Valdez, to serve as the official coronation ceremony.

However, this new consciousness is not a monolithic, narrow view. It has many component parts, most of which had their origins in the 1970s and 1980s and are often in flux, depending on the nature of a specific issue.

Still, the collective weight of these strains of consciousness often add up to a new law or new regulation that dictates the course of business decisions.

Those business decisions will invariably translate into more dollars. Spending for pollution control and regulation in the United States by government, industry, and households, now approaches 2% of the gross national product. If nothing else, the New Environmental Consciousness means that number will surely rise.

Thus it becomes instructive to understand the components of that mentality and especially to understand the tactics and methodologies of those in the vanguard or promoting that new consciousness.

The Nimby Syndrome

. . . .

"Not in my back yard, you don't!"

No doubt the *Not-in-my-back-yard* (Nimby) syndrome has been around in one form or another since the dawn of humankind, having its roots in some prehistoric version of the territorial imperative—anthropologist Robert Ardrey's term for primates' instinctive need to mark off and defend part of their physical environment for a sense of security.

The Nimby syndrome covers a wide range of human activity. Residents of an area might object to a proposal for a new freeway, prison, or landfill; all useful things for society at large but each carrying a varying degree of impact for its neighbors.

The Nimby syndrome probably won its widest validation with the so-called disasters at Love Canal and Three Mile Island in the 1970s. To this day, there is no proof that either created a real health hazard, through presumed radiation exposure or chemical exposure. Yet the two are often linked in Americans' minds with their analogous tragedies at Chernobyl and Bhopal in a kind of "It could happen here" scenario.

Although the facts would argue otherwise, the Nimby syndrome usually operates on a diet of fear with the main course a worst-case scenario. When an "It can't happen here" event such as the Exxon Valdez *does* happen, then it escalates the Nimby syndrome into dogma—an article of faith, no matter what the statistical odds of a repeat.

Nowhere is the Nimby credo more dogmatic than in California. There was no lasting damage from the Santa Barbara spill—to the contrary, fishing and tourism have flourished in the area since the spill—and there has not been a significant spill as a result of drilling-production operations in California waters since. It is an article of faith among the populace there that more catastrophic spills resulting from exploration and production operations off California are inevitable.

The fact that the February 1990 American Trader spill off Huntington Beach, California, was a tanker accident did nothing to change that view. As with the Exxon Valdez, the linkage of the American Trader spill with leasing and drilling-production off California was as immediate as it was false. Every barrel of oil produced off California and brought via pipeline to shore is one not transported by tankers—with their much greater risk of an oil spill. The lie is that shutting down leasing and drilling off California reduces the overall risk of an offshore oil spill when in fact it increases that risk by mandating more tanker-borne imports.

During much of the debate over leasing federal lands off California—locked up by congressional bans since the mid-1980s—the petroleum industry effectively knocked down most opponents' arguments against leasing on environmental grounds. Industry officials cited a sparkling operations record and studies that showed no significant ill-effect from drilling and production operations on the air, water, or marine life. Eventually, opponents were reduced to invoking the ultimate expression of the Nimby credo as only a real estate-obsessed Californian could put it: "Harmful to the environment or not, I don't even want to see it in my back yard, and my back yard extends at least 12 mi offshore."

Absurd? Of course, unless you're one of the well-to-do coastal residents (and it's pretty tough to be the latter without also being the former) of California. The concept of viewshed in real estate law has been stretched to its ultimate in California. Among the arguments raised against drilling was the notion that rigs or platforms stationed even temporarily within sight of someone's vista of the ocean constituted a sufficiently unaesthetic industrial intrusion so as to threaten property values.

Imagine, having to lop a few thousand dollars from the asking price of some million dollar Malibu manse, weighed against mere national interests: energy security, budget and trade deficits, jobs, etc.

California offshore operators have had to contend with a once-booming province that was the hottest U.S. play in the early 1980s fizzling into a series of hamstrung development projects. Early estimates of what Offshore California would produce as a result mainly of discoveries in the late 1970s and early 1980s ranged to as much as 500,000 B/D of oil in the mid-1990s. Of those discoveries, only one made after 1980—Unocal Corp.'s Point Pedernales field off Vandenberg Air Force base—is producing as this book goes to press. Three platforms, part of a $2 billion project to develop Point Arguello field, still lay idle after three years of being ensnarled in permitting delays at the local and state level. That is despite its rank as the biggest field ever found in U.S. waters and the herculean efforts of Chevron Corp. in accommodating environmental concerns by spending hundreds of millions of dollars on mitigation and in earning good will among the locals through a grassroots community outreach program.

The culprit: the Exxon Valdez. Although Chevron had won its final permit from Santa Barbara County to ship Point Arguello crude to its Los Angeles refinery via tanker, the California Coastal Commission overturned the county permit on an appeal by GOO the League of Women Voters. The rationale: fear of a spill, although the Chevron crude being tankered would merely displace the same volume of Alaskan oil being tankered through the same area.

Again, facts and science fell by the wayside. The perception was an increased risk of oil spill. With the Nimby syndrome, there is an immutable law: *Perception = Reality = Political Action.*

The Zero Risk Game

· ·

The most critical component of that perception is risk, specifically the degree of risk.

The science of risk management is sufficiently young so as to be scarcely considered a science. No manager is going to work on the assumption that any operation will carry a no-risk guarantee. And yet that is the direction public policy increasingly is heading.

Elizabeth Whelan, founder and director of the American Council on Science and Health, writes in Chevron World[1]:

> Prudent and constant vigilance is necessary to prevent health-threatening exposure to potentially hazardous substances. There is no debate about this. But many Americans are unable to distinguish between proper regulation versus outright bans of useful chemicals that don't threaten health.
>
> The most critical of toxicological adages seems lost on many: "Only the dose makes the poison." In other words, if any trace of a toxic—no matter how infinitesimal—is detected in something we eat, drink, or breathe, a growing number of people want it eliminated.
>
> This has translated into a mandate that regulations should ensure that we live in a world where there's "zero risk at any cost." That is impossible. Scientists have another word for a zero risk world—a vacuum, and you can't live in one for long.[1]

Dr. Whelan cites as the most extreme example of this new thinking California's Proposition 65.

Prop 65 is an initiative that California voters passed in 1986 that essentially forbids the discharge into or near drinking water any detectable amount of a substance thought to be carcinogenic or teratogenic (able to cause reproductive illnesses or birth defects).

What is especially nettling about Prop 65 for the petroleum industry is the almost whimsical nature of its implementation. Initially, the list of substances to be covered by the measure were a few dozen that have been established as carcinogens or teratogens. Prop 65 activists—led by Hollywood's glitterati—lobbied the governor to increase the number of banned substances to more than 300 substances *suspected* of causing cancer or reproductive/birth defects.

Much of the measure's impact has been diluted thus far by confusion over what the standards for detectable amounts are and how it is to be implemented. One especially chilling aspect of Prop 65 is the so-called "bounty hunter" provision. It requires the state to pursue any claim by anyone that someone is discharging said banned substances into water and offers a reward out of a percentage of any fines collected. This could lead to a spate of nuisance actions by disgruntled former employees, pranksters, or simply serve as a venue for economic terrorism by environmental activists.

Thus far, there has been little actual economic damage wrought by Prop 65. However, the one visible sign of the measure's effects is the ubiquitous spread of Prop 65 warning signs. One year after a chemical goes on the Prop 65 list, businesses are required to give anyone exposed to the chemical clear and reasonable warning. The list exempts food, drug, and other consumer products regulated under state and federal safety laws.

There is hardly a business establishment or worksite in California that does not have the posted sign reading:

"WARNING: Detectable amounts of chemicals known to the state to cause cancer, birth defects, or other reproductive harm may be found in and around this facility."

Perhaps this kind of saturation becomes meaningless after awhile, as people become blasé over "Big C" alarms as being nothing more than an updated version of the Boy Who Cried Wolf.

Or perhaps, on a subliminal level, this ubiquity serves to seep into people's consciousness that any level of exposure to these chemicals is a threat, inculcating a sort of chemophobia that helps pave the way for mandated zero-discharge solutions on almost all waste and emissions.

58

Dr. Whelan notes that there has never been any scientific evidence to claims that the substances covered by Prop 65 have harmed anyone and there has never been a proven link between what happens in lab tests on animals and what happens to humans.

> The odds of cancer in laboratory animal tests have always been stacked in favor of the "Big C." Animals known to be extremely sensitive to particular chemicals are chosen, and the highest nonlethal dose is used. In fact, there is little connection between a small rodent exposed, for instance, to the saccharin equivalent of 1,800 bottles of soda pop a day and a human's daily intake.[1]

Nevertheless, Dr. Whelan argues, "toxic terrorists" are exploiting such tests to prey on the American public's fears.

> We must realize that toxins are all around us in the natural environment and they always have been . . . And with our technological ability now to detect substances in the parts per trillion, we are approaching the point where we can identify some trace of a toxic molecule in every breath we inhale.

> This is nothing to be alarmed about. However, it is critical that we learn to assess the difference between a harmful toxic amount and a harmless amount.[2]

It is difficult to sort out that difference, however, when the "Toxic Terrorists" take to the nation's air waves to declare again that the sky is falling.

There are considerable risks in attempting to achieve a zero-risk society.

Amoco Corp. Chairman Richard Morrow told the Houston Club Forum in 1989 that a risk-free environment is an impossible goal, and "the attempt to reach it needlessly wastes precious resources that might otherwise be utilized to remove more immediate and genuine threats to health and safety."

"Risk and danger are alien obstacles to overcome," Morrow said. "They are part and parcel of the human condition—the reality of an imperfect world.

"Instead of a preoccupation with freedom from all risk, we should concentrate our effort on achieving maximum practical levels of environmental safety."

There is also the consideration of economic risk. The increased focus on a zero-risk environment imposes severe economic burdens on an American business community, putting it at much greater risk in competing globally—where it already is suffering.

Just how extreme is the environmental lobby when it comes to weighing risk? Here's an example:

Responding to the hysteria over pesticides in late 1989, the Bush administration proposed a pesticide monitoring plan that include a provision for the EPA to take into consideration the value of a chemical when deciding whether to ban it. Most chemicals face the old standard of being banned in the event they cause one case of cancer in a million over a lifetime. However, chemicals deemed valuable and irreplaceable would face a standard of one in 1,000. Considering the usual rule of thumb where a chemical gets banned "at the drop of a rat," it would seem a reasonable compromise, right?

Here's the Natural Resources Defense Council response to that proposal, as carried in press reports of the day:

"It's a tremendous disappointment. Allowing the EPA to condone continued use of a chemical whenever the benefits outweigh the risks is absolutely anathema to the environmental community."

If that logic seems straight out of Lewis Carroll, then you could be safe in assuming that the U.S. environmental agenda is truly directed by a "Mad Hatter."

In other words, there is *no* acceptable level of risk as far as the environmental lobby is concerned. It isn't just bad science. One must understand the larger agenda of the environmental movement to comprehend the rationale (for want of a better word) of that stance.

A Digression on Definitions

. . . .

It would be too easy to lapse into a discussion of political conspiracies, but I suspect the political philosophies of those in the environmental movement are secondary to the discussion, however much they might be in accord with each other.

First, there is the question of what is an environmentalist, anyway. That has always been a misnomer. Dictionary definitions simply refer to the word "environment" as "surroundings," a "complex of external physical conditions influencing the growth and development of organisms," or "the complex of social and cultural conditions affecting the nature of an individual or community."

Nothing in there about "natural" versus "manmade," or "organic" versus "synthetic." It's like the distinction between opposing sides on the abortion issue. Instead of simply and clearly stating that one is anti-abortion or pro-abortion, it adds the perception of moral weight to one's position by claiming to be pro-life or pro-choice. Thus, by claiming to be an environmentalist, one automatically takes the moral high ground, consigning those who would oppose them as being "anti-environmentalist," or someone who actively seeks to despoil his surroundings. Ultimately, being anti-environmentalist thus takes on the cachet of being bent on one's own destruction. And so being environmentalist implies saving ourselves from our own destruction, taking on a quasireligious hue.

That missionary-saving-our-souls angle is a part of the environmentalist attitude that cannot be underestimated. It is instructive to recognize that the modern environmental movement had its earliest roots in the writings of Henry David Thoreau and John Muir, who both professed a vague sort of pantheism expressing itself through the glories of nature. It is equally helpful to read through some of the newsletters and books and pamphlets urging action on the environment. Everywhere, Man is the villain of the piece. Pristine wilderness is Eden, Heaven, salvation; the Hand of Man in any shape in that wilderness is sacrilege, the Serpent in the Garden. There is, it seems, no middle ground. This attitude is a fundamental part of the environmentalist mentality, whether it entails the sturdy Jehovah's Witness-type proselytizing of the shopping mall pamphleteers and door-to-door canvassers to the environmental jihad-style terror tactics of some of the extremists, with their life-threatening sabotage efforts.

This might suggest why the environmental movement has not co-opted the word "ecologist" as their official label. It may well be because ecology truly is a science, the science of the relationships between organisms and their surroundings, and there is little room in Environmental Holy Writ for science, much less economics.

I would like to propose a more suitable label for those who dub themselves "environmentalists" in the strictest activist sense: Neodruids.

(That's with apologies to one of the finest living American writers, John McPhee, whose brilliant book *Encounters with the Archdruid* is the best study of the conflict between environmentalists and developers and the keenest analysis of the environmentalist mentality. His "Archdruid" is David Brower, executive director of the Sierra Club, whom he likens to the evangelist Billy Graham.)

The Druids were an ancient order of priests/magicians who wielded great influence over the Celtic peoples of Britain, Ireland, and Gaul. They worshipped the oak tree and mistletoe, offering up living sacrifices in secret ceremonies in oak groves.

The Neodruids, then, are those who ascribe to a worship of nature and would sacrifice human interests in pursuit of salvation, achieved by preserving untouched the Holy Temple of Nature.

A harsh assessment? Think about terms such as "rapacious greed," "rape and pillage," "wanton disregard for environmental values" that have peppered the environmentalists' descriptions of economic activity versus environmental values.

That may describe the mentality of the typical grassroots community activist, but for the professional lobbyists and lawyers and organizational officers advancing the environmentalists' agenda in Washington and the state capitals, that view must be leavened with the recognition that there is a more personal, self-serving agenda as well.

The only way that environmental groups can survive, much less thrive, is to attract grants from foundations, government grants, or individual and corporate contributions. Their ability to attract those revenues are directly linked to their ability to attract attention, chiefly by making news. At the same time, a daily press corps that is all too willing to suspend disbelief become unwitting allies in that cause by not questioning the environmentalists' assertions too closely. Is it not time that the press cast the same kind of close eye on these so-called public interest groups, as they did during the recent television evangelist scandals?

Mixing Alar Apples and Chilean Grapes; or, the Ineffectiveness of Science against a Media Rollout

Hysteria fed by a media barrage nearly destroyed the U.S. apple industry and Chile's fruit export industry in 1988. Was the fear justified? The record suggests that the American public and these two industries were victimized needlessly.

How the Alar scare came about and why it was such a success for those promoting that fear is illustrative for the petroleum industry.

The Alar scare was the result of a carefully orchestrated media campaign staged by a political publicist, David Fenton of Fenton Communications, in concert with the NRDC. That was detailed in a memo that Fenton sent to "interested parties" in the press.

NRDC was about to publish a report, "Intolerable Risk: Pesticides in Our Children's Food" in fall 1988 when it hired Fenton to coordinate the media campaign for the report. About the same time, the renowned actress Meryl Streep contacted NRDC to ask if she could help with some environmental projects. She read the report's preliminary results and agreed to act as a spokeswoman for the project.

Under an agreement with NRDC, one week after the study was to be released Streep and others were to announce the formation of NRDC's new project, Mothers and Others for Pesticide Limits, a lobby group to seek changes in pesticide regulation and push for pesticide-free produce.

Meanwhile, Fenton struck an agreement with the CBS program "60 Minutes" to break the story and arranged interviews months in advance with the major women's magazines and top daytime television talk and news shows.

The day after "60 Minutes" broke the story, NRDC held a massive news conference in Washington, D.C., as well as 12 local news conferences around the country to release the report.

A little more than a week later, Streep held a Washington news conference and announced the creation of the NRDC lobby group. Correspondingly, all the major print and electronic media jumped on that story as well.

The two-pronged approach, boosted especially by the capper of star quality, helped sustain the Alar issue until it began to take on a life of its own. The fear spread, and soon school systems were banning apples, and apple growers' sales were plummeting.

The EPA, U.S. Department of Agriculture, and Food and Drug Administration issued a joint statement assuring the public that apples were safe, with the NRDC issuing the usual vague caveats about what levels of toxicity were acceptable.

Shortly thereafter, two Chilean grapes laced with cyanide were discovered. That pushed the Alar scare into a full-blown "Big Issue" on food safety destined for the newsmagazine covers.

The government and the food growers trotted out scientist after scientist to refute the fearmongers' claims.

Little attention was paid to the fact that almost all plants contain natural pesticides, toxins that are a defense against insects, fungi, and predators. Most tests for carcinogenicity involve synthetic chemical compounds. Seldom are the natural toxins eaten in fruits and vegetables every day subjected to such tests. When they are, the result can be surprising.

In tests of natural and synthetic toxins Dr. Bruce Ames, Chairman of the Department of Biochemistry at the University of California-Berkeley, concluded that the proportion of positive tests is about as high for natural pesticides as for synthetic chemicals.

Since more than 99.99% of the pesticides humans ingest are "nature's pesticides," our diet is likely to be very high in natural carcinogens. Their concentration is usually in parts per thousand or more versus the parts per billion that is the usual yardstick of measurement for synthetic pesticide residues or water pollution, Ames found.

All of that notwithstanding, the typical mom could only relate to the television commercials depicting a radiant Streep and her young daughter dutifully scrubbing vegetables with soap.

A similar tactic worked for Prop 65: Jane Fonda produced a film showing the deformed children of mothers supposedly poisoned during pregnancy by toxins in the drinking water.

Lest one doubt the efficacy of a big media rollout complete with celebrities, consider a more recent effort and the chill it has sent through the petroleum, chemical, and other industries.

In summer 1989, Henry A. Waxman, chairman of the House subcommittee on health and the environment staged a press conference and proffered his "dirty dozen" list of the nation's worst toxic polluters, specifically citing ranking industrial installations by degree of cancer risk.

The list was derived from data gathered under the Emergency Planning and Community Right-to-Know law—Title III of the Superfund law. As of July 1, 1989, plants in the United States are required to disclose their annual emissions of toxic chemicals.

Americans love lists, as witnessed by collegiate sports polls and various top 10 tallies throughout the arts. Compiling a list of the worst polluters is going to be more effective than simply using generalized data about the amount of toxic pollution or even dealing with the nature of the threat, if any. It doesn't matter whether or not the claims are valid.

At the top of the Waxman list (a California representative who is one of the environmental lobby's key proxies in Congress) were Texaco Inc. and Unocal Corp., both of whom strongly took issue with the claims.

The allegations that its La Mirada, California, polymers plant is one of the worst toxic polluters are simply not true, Unocal Chairman Richard Stegemeier said.

Unocal noted that the EPA risk assessment Waxman used in compiling the list is not based upon the plant's actual record, but upon a computer model for butadiene plants. In addition, the EPA discredited use of the data, and the agency's administrator William K. Reilly—former head of the National Wildlife Federation—said in a letter to Waxman that "the facility specific risk estimates are not accurate determinants of local public health hazards . . . these data are not to be relied on as credible estimates of risk for individual sources."

EPA later said the figures were never used to reflect risks in the real world and that any suggestions that those data reflect such risks was "entirely erroneous" and "highly irresponsible."

Texaco also complained about Waxman listing its Port Neches, Texas, chemical plant as one of the worst toxic air polluters, contending the congressman offered "unsubstantiated estimates of cancer risk that go far beyond the caveats and restrictions which the EPA brought to his attention regarding the reliability of such data."

The company also said that Waxman's claims were based on plant configurations no longer in existence at the plant and are refuted by tests on actual emissions at the plant. In addition, the conclusions disregard the millions of dollars for environmental expenditures Texaco made at the plant in recent years to improve air quality.

Think that took the wind out of the environmental lobby's sails?

Not two months later, the National Wildlife Federation staged a big media rollout for its list of the "Toxic 500," the industrial facilities that account for most of the toxic chemical releases reported to the government. Although the toxic chemical waste generated (notice the different coloration the word "releases" provides?) was perfectly legal, the Big C was invoked again and the media wolfpack was off and running again.

Marketing "Green"

. . . .

The media overkill as practiced by the environmental lobby on purported crises is buttressed by a corresponding volume of hype as Madison Avenue essays its own brand of the "Greening of America."

(If you think of this effort in terms of dollar bills, it might more appropriately be called the "De-greening of America.")

Madison Avenue recognizes a fundamental change in American attitudes, and it already is skewing its advertising message to further enshrine the environmental ethos in American popular culture.

Corporate America is recognizing that change as well, and it is growing increasingly skittish about a resurgent militant consumerism.

In a *Business Week*/Harris poll published in the May 29, 1989, issue of the magazine, results showed Americans held deep misgivings about business ethics.

In response to a question asked of those surveyed about how far business would go in order to boost profits, 47% said business would resort to harming the environment; 42%, put workers' health and safety at risk; 38%, endanger public health; and 37%, sell unsafe products.

Further, almost three-fourths of those surveyed said environmental pressure groups do more good than harm.

The survey also showed that Americans are more willing to take action against business whose behavior is deemed unacceptable. Protesting a company's actions would range from a boycott of products (76%) to writing letters to a CEO (56%) to contributing money to groups opposing the company (21%) to picketing (14%) to sabotage (2%). The poll also clearly showed the majority of Americans believes that public activism can change corporate behavior.

Those kinds of signals are tough to ignore, and marketing and advertising specialists are already responding. Witness the proliferation of biodegradable and recyclable products in stores in just the past year. One of the savviest of American retailers, Sam Walton of Wal-Mart fame, has launched an effort to tag products in his stores that are deemed to carry some environmental benefit. This echoes his "Buy America" effort that began a few years ago, which Madison Avenue has deemed a big success. You just can't get any more Mainstream America than Wal-Mart.

That trend is likely to grow. A Canadian food distributor and grocery store chain last year introduced a "Green" line of environmentally benign products backed by a huge advertising campaign. Among 100 items are biodegradable diapers, bathroom tissue from recycled paper, and foam plates made without using chlorofluorocarbon as blowing agents.

Simply put, "green" sells. Nowhere is that more evident than in the debate over garbage. Although there is strong evidence to point to biodegradability as something of a sham, companies are pursuing marketing of biodegradable products vigorously. The sham can be seen in studies that, under the right conditions, even food can survive decades buried in a landfill. So much for decrying plastic as the chief nonbiodegradable culprit.

However, even plastics producers are introducing biodegradable plastics, despite the evidence, in responding to consumer demands.

The shift in the American consumer's attitude is fundamental enough to overcome some deep-seated phobias. Consumer fear over pesticides has led to changes in product development and marketing strategy by the manufacturer of Raid insect sprays. Psychological profiles of consumers, notably housewives, have shown they had a deep-seated need to watch cockroaches sprayed with a bugkiller writhe and spin in circles before dying. However, makers of insect control products recently have shifted marketing strategies to focus more on baits and traps with an idea more toward control versus total annihilation of the pests. If anything, the currently fashionable reasoning goes, roaches, with their roles in decomposition, have environmental benefits versus the "deadly" pesticides.

Wall Street is taking notice of environmental marketing as well. In 1989, a group of 100 institutional investors drew up a vaguely worded code of conduct for businesses regarding environmental concerns, dubbed the "Valdez principles." The implication was that companies not adhering to the Valdez principles in the course of their operations would not attract these institutions' investments in their stock.

Similarly, the Teachers Insurance and Annuity Association-College Retirement Equities Fund introduced two investment funds for educational institutions and their employees in March 1990. One of those, the CREF Social Choice Account, would seek to provide returns that reflect the broad performance of the market while investing in companies "that conduct their activities in a manner that recognizes acceptable standards for addressing specified social concerns."

Among those concerns, which include ties to South Africa and production of nuclear energy, guns, alcoholic beverages, and tobacco, are environmental issues, which are "expected to be an important consideration in the conduct of the CREF Social Choice Account's investment policy. Practical criteria will be sought permitting systematic exclusion of companies that fail to adhere to sound environmental policies and practices."

How the "Greens" Market Themselves

• •

The environmental groups are just as savvy about Madison Avenue and Wall Street as they are about Washington, D.C.

The Exxon Valdez not only had a galvanizing effect on the environmental organizations' causes, it provided a tremendous boost to marketing efforts, including subscriptions and memberships.

Environmental groups were quick to capitalize on the tragedy, sending out a flurry of solicitations for financial support and new members.

In 1988–89, contributions to the National Audubon Society jumped more than 50%, and the rate of growth in Wilderness Society membership increased by more than 100%.

The National Wildlife Federation, with membership of 5.1 million in spring 1988, created an Alaska Fund right after the spill asking for contributions to research the spill's effects and lobby against drilling in the Arctic National Wildlife Refuge. It received more than $400,000 in the first two months after the spill. Between May 1988 and May 1989, NWF boosted its membership by 700,000, to 5.8 million.

From the start of 1988 to midyear 1989, the Environmental Defense Fund's membership more than doubled to more than 100,000.

Between fall 1988 and spring 1989, Greenpeace added members at the rate of 50,000 per month, growing to 1.35 million.

Again, a canny use of the media helped the environmental groups' efforts.

For example, NWF, Audubon, and the Sierra Club a month after the spill announced plans to sponsor a National Alaska Wildlife Memorial Day, set for May 7, "to acknowledge the catastrophic effects of the oil spill on the world's animals and the ecosystem."

With that, the organizations sent materials to schools around the country to participate in a national day of mourning for Alaskan wildlife.

The pleas for contributions, which often were accompanied by preprinted telegrams to mail to Exxon and President Bush demanding action, were hardly subtle: "I'm asking you to act now or say goodbye to the Alaskan wilderness." "I'm mad as hell, and I'll bet you are, too!" "Like vultures, they continue to circle Alaska. The tragedy of Prince William Sound apparently means nothing to the oil developers."

The theme was a common one: The threat That This Will Happen Again was invoked as justification for seeking a halt to plans for leasing and drilling in ANWR, to block leasing offshore Alaska, California, and the East Coast, to pursue energy conservation and alternate fuels subsidies.

Some environmental groups even carried their righteous indignation so far as to actually refuse funds from Exxon (of course, putting out a press release to that effect).

The enmity aimed at Exxon resulted in an interesting brouhaha in fall 1989, when Exxon resigned from the NWF's corporate advisory panel, saying the organization had been unfairly critical of the company's handling of the Exxon Valdez spill.

The issue came to light when NWF made public an exchange of letters between Exxon's environmental coordinator Ray Campion and NWF's President Jay Hair. Exxon had been a charter member of the NWF Corporate Conservation Council.

"Recent public actions by you regarding the Valdez oil spill have failed to demonstrate any sense of objectivity or fairness," Campion wrote Hair. "In our judgment, the federation's role in the council has not met our expectations in terms of seeking balanced solutions to the nation's environmental problems."

Hair returned from Alaska in September, and, in a widely publicized stunt, brought a pile of oil-stained rocks to Washington to distribute to each congressman to demonstrate what he considered Exxon's ineffectiveness at cleaning up the spill. In addition, NWF was one of a number of plaintiffs in a multibillion dollar lawsuit against Exxon seeking full restitution for environmental recovery costs.

"I regret that we did not meet your 'expectations,' but did you really think your membership on our Corporate Conservation Council would buy Exxon immunity from the National Wildlife Federation's response to such a massive and poorly managed environmental disaster," Hair replied. "The fact that Exxon has been judged a corporate pariah in the court of public opinion has little or nothing to do with the National Wildlife Federation. That distinction has been clearly earned by Exxon's leadership."

The federation kept the heat on Exxon for months to come.

Shortly after the exchange of letters, NWF put out a press release condemning the federal government's draft plan for assessing damage to Prince William Sound, notably a provision that could permit Exxon to conduct some aspects of the damage assessment.

"Letting Exxon determine how badly the sound is damaged is like letting Count Dracula take inventory at the blood bank, it said."

NWF claimed the government's plan would allow Exxon to escape the full cost of cleaning up the sound.

A few weeks after that statement, Hair went to Alaska to conduct "citizen hearings" into the spill and its effects and took the opportunity to lash Exxon again for "isolating itself from concerns about the environment," having "bunker mentality," and being "arrogant."

That was a high-profile, extended barrage aimed at Exxon. But many environmental groups wanted the linkage to ecological destruction not just to Exxon or even the oil industry, but to the use of oil itself. The message then became: The *real* problem is excessive U.S. dependence on oil. In short, we won't have to drill off our sensitive coasts (or presumably, tanker in any more imports) if we just add a littler insulation here, lower the speed limit and the thermostat, and install solar water heaters. And there are ancillary messages: Using oil contributes to the global warming trend, worsens air and water quality, creates plastics and other petrochemical products that harm the environment either through process or end product disposal.

Environmentalist Tactics in Washington

Environmental groups also are marketing themselves differently in Washington.

The environmentalists of the late 1960s and early 1970s have traded their jeans and boots for three-piece suits and Italian loafers. They have become extremely sophisticated at lobbying efforts, maintaining computerized databases showing the environmental voting records of congressmen, hiring Harvard trained lawyers, and organizing local grassroots campaigns to pressure representatives at home.

Beyond the lobbying, environmentalist groups are infiltrating the infrastructure of the federal government at every level, not just highly visible posts like EPA administrator.

A coalition of more than 20 environmental groups formed the Project Blueprint Talent Bank in Washington in 1988. It represents the heart of a drive to encourage as many environmentalists as possible to apply for jobs within the executive branch of government, in political and career positions.

The rationale is that many environmental issues succeed or fail depending upon how they are implemented. By creating a cadre of environmentalists within the federal bureaucracy, the environmentalist groups can effectively guide environmental regulation and manage the government's environmental programs as they see fit.

Sierra Club Chairman Michael McCloskey, writing in the *Environmental Talent Bank's Guide to Federal Jobs,* called for environmentalists to seek a broad range of government positions to offset the prospect of an unsympathetic bureaucracy that can sabotage environmental programs:

> We need them holding political appointments as senior managers and policy experts. We need them as regional directors. We need them in mid-level professional positions, where so much of the work gets done. And we need them at the beginning level, where people can build careers and provide for the future.
>
> We cannot leave these agencies and departments to those who are indifferent or who have other agendas.[3]

Environmentalist Strategies

· ·

It is crucial to understand some of the strategies and tactics of the environmental lobby in Washington.

Although there was a perception that the environmental movement was in retreat before the onslaught of the Reagan deregulation juggernaut, that was probably more a sense of a broader rejection by the public of the knee-jerk environmental extremism that dominated so much government policy in the 1970s.

That change in the public's attitude probably paved the way for some key advances for the petroleum industry in the 1980s: a stronger U.S. leasing program on the Outer Continental Shelf; completion of oil price decontrol; natural gas deregulation moves; greater emphasis on applying cost-benefit yardsticks to new environmental regulations; rollback of some conservation measures; and a generally stronger economy boosting oil and gas demand.

However, far from being in retreat in the 1980s, the environmental lobby prospered, thanks largely to one man: former Interior Sec. James G. Watt. It was Watt's stormy tenure at Interior that may have planted the first real seed for the environmental lobby's renaissance that flowered at the close of the last decade.

Watt's adversarial relationship with the environmental lobby and its allies in the Congress energized the movement as nothing else had since the Santa Barbara spill. He accomplished much on federal lands access, notably on the OCS lease sale program.

However, Watt—especially with his controversial personal style and unconventional views—served as a lightning rod for the environmental lobby and congressional critics of the Reagan administration. The environmental lobby mounted in the first two years of the first Reagan term a systematic campaign aimed at undermining and later seeking the ouster of Watt.

The outpouring of support in response to that effort was clear. The Sierra Club, for example, saw its membership rolls jump by more than 25% to more than 245,000 in the first nine months of 1981. Emboldened by success, the environmentalist lobby sought to expand its attack on other Reagan administration officials, notably EPA Administrator Ann Gorsuch, Department of Energy Sec. James Edwards, and Department of Agriculture Sec. John Block.

The environmental lobby grew more politically sophisticated in the 1980s. Environmental groups began organizing political action committees (PACs) to drum up support for tougher environmental policies. The PACs worked hand-in-hand with the League of Conservation Voters to implement electoral strategies. By 1982, there were 14 states with environmental PACs actively working to oust or elect state and congressional candidates. The League of Conservation voters spent almost a half million dollars in a door-to-door canvassing effort that reached more than 400,000 people in the 1980 election. Such efforts were instrumental in securing the election of several allies in Congress, governorships, and state houses that year.

Further, environmentalists in the 1980s began efforts to collaborate with other special interest groups, such as the American Cancer Society and the March of Dimes. The idea was to enlist the aid of health charities, and thus tap their huge lobbying funds to battle budget cut proposals on environmental regulation. Similar overtures had successfully been extended to some of the major labor unions—despite the overwhelming evidence that so many aspects of the environmentalist agenda would result in the loss of thousands of American industrial jobs.

In a 1982 study commissioned for the Republican Study Committee, Tim Peckinpaugh provided a compelling analysis of the environmental lobby and its tactics. Peckinpaugh contended the environmental movement had expanded its objective to encompass a larger sociopolitical agenda.

He cited Dr. H. Peter Metzger, manager of public affairs planning at Public Service Co. of Colorado, who described environmental extremists as "coercive utopians" who have a hidden agenda—not merely seeking a clean and safe environment, but also striving for "some vague political goal, designed to come about by stopping energy production as we know it."

A typical example of this rationale is found in a May 25, 1989, press release, announcing a call to revolution by the Fossil Fuels Policy Action Institute (FFPA):

"FFPA wants to focus the issue of survival more broadly than merely cleaning up the environment," the release said.

"The nation is beginning to adopt a more broadly based and widely supported commitment to a new social ethic that will conserve the world's precious resources . . . the task and our predicament call for nothing less than a revolution."

As FFPA asserts, that calls for a conservation revolution:

"As truly concerned citizens, wishing to prevent future Exxon Valdezes, we should take concrete action to reduce massively our dependence on crude oil and fossil fuels in general."

Peckinpaugh also punctures some myths about the environmental movement's presumed altruism, citing its alliance with the promotion of "soft" energy technologies and thus evidence of business and employment opportunities therein.

He cites authorities who have concluded environmentalists are fundamentally self-interested and act as self-interested contenders for a publicly controlled resource.

Peckinpaugh contends that environmentalists tend to be affluent, upper-middle-class who are largely insulated from the consequences of stagnant resource development and economic growth:

> Furthermore, many members of the leisure class stand to gain the most from complete preservation of scenic refuges because only they have the time and money to frequent such retreats . . . The recreation and scenic value of unharvested forests, for instance, outweighs the value of economical timber provided you are a member of the affluent "wine and cheese belt," and not an employee of a lumber mill or a member of the lower-middle-class seeking to purchase a home.

> Environmentalists thus are self-motivated to thwart economic development through resource preservation because only they garner the benefits of extremist environmental protection, and only they are isolated from the harmful consequences of sluggish economic activity.[4]

The environmental lobby, in addition to its skill at manipulating the political, legal, and regulatory processes, is equally skilled at extracting funds from the American taxpayer to subsidize its agenda.

Many environmental groups in the 1970s and 1980s won grants and contracts from federal agencies, although the trough dried to more of a trickle later, during the Reagan years. Further, some environmental organizations retain some sort of tax exempt status, receiving an indirect federal subsidy through income tax deductions.

It is hardly just Uncle Sam footing the bill for so much environmental lobbying. Corporate America has long been contributing money to some of the environmental groups, including those that have worked hard to block the contributors' own projects. U.S. petroleum companies are among the biggest givers to those who would bite the hand that feeds them.

A Note on Science

. . . .

This is a good place for a brief digression on the state of the science surrounding some environmental concerns. It may very well be that the 1990s will be remembered not as the Decade of the Environment but as the Decade of the Chicken Little Syndrome.

Contrary to what many in the environmental lobby may say, and its concomitant "validation" by unquestioning media repetition, there is no true scientific consensus on whether such threats as global warming from the greenhouse effect, acid rain, and ozone depletion are *even real,* much less a consensus as to how dire said threats are. There is even less of a consensus as to the best solutions from a purely environmental standpoint, much less a consensus on the best, cost-effective environmental solutions.

Much of the debate in the early 1990s over these issues—especially global warming—will begin to take on the quality of "How many angels can dance on the head of a pin?" because there is much that is fuzzy in global climatological models and speculative about atmospheric chemical reactions. In short, the

74

number of speculations about such threats and proposed solutions to them will rise in direct proportion to the amount of federal grant money made available, with science probably the chief victim in this assault of theory.

Nevertheless, it behooves the petroleum industry to remember the fundamental rule about the effectiveness of the Chicken Little Syndrome: *Perception is reality.*

At this point, the petroleum industry's strategies on environmental issues should not be based on a foundation of pure science. That may prove difficult for the scores of those schooled in the scientific method who make the industry's business decisions these days. Science is not irrelevant, and industry should not shrink from holding environmentalists' feet to the fire when it comes to shoddy science.

But being right is not enough. Even Galileo was forced to recant. Look at the cyclamates scare. A whole industry was disrupted when study after study showed the American public risked getting cancer by ingesting cyclamates. Although there were efforts to refute some of the early scare claims, noting that one would have to drink a bathtub full of the stuff every day for 20 years to equal the cancer risk of a cigarette smoker, the prevailing "wisdom" held, and cyclamates have remained under a cloud all these years. However, studies emerged late in 1989 that seemed to prove conclusively, once and for all, that there is no significant cancer risk in ingesting cyclamates. It did not matter that the cyclamates producers were right. The perception created among the public is what held sway.

Industry should act accordingly. Global warming may be a real threat in the middle of the next century. Or it may prove to be another Comet Kohoutek. Petroleum companies should take into consideration what those worst-case scenarios are and plan accordingly, while building their own case to refute the popular assumptions. In other words, do something about the current level of emissions that are thought to contribute to the greenhouse effect now, and thus avoid the replacement of your product or commodity with a substitute mandated by government.

The apocalyptics have the momentum on their side, just as they did shortly after the Santa Barbara spill. They will probably maintain that advantage until economic or energy or other geopolitical crises distract, as they did in the latter 1970s. By that time, there will no doubt have been sweeping changes in how the petroleum industry does business, irrespective of the quality of science underlying those changes. Remember that the principal guiding force of environmentalism in the United States is not science, or even the public welfare, but *politics, politics, politics.*

References

· ·

1. Dr. Elizabeth F. Whelan, "Zero Risk," Chevron World, Spring/Summer 1988.
2. Ibid.
3. Michael McCloskey, "From the Outside," Environmental Talent Bank's Guide to Federal Jobs, Renew America Report, September-October 1988, 4.
4. Tim Peckinpaugh, "The Specter of Environmentalism: the Threat of Environmental Groups," a special report to the Republican Study Committee, U.S. House of Representatives, February 12, 1982.

4

GOVERNMENT LEADS THE WAY

Washington's Role

• •

The role of the federal government during the (first?) term of the Bush adminis-tration will be crucial in determining the direction of environmental policy in the United States for the 1990s and beyond.

That was certainly heightened by the dramatic events of 1989–90 that swept the USSR and its former satellites. There has been much speculation over a massive rollback in U.S. military spending with the Soviet military threat fading. In fact, no sooner had the Berlin Wall been breached than congressmen and special interest groups—with the environmental lobby in the forefront—already had lined up at the trough to spend a "peace dividend" expected to result from the changing geopolitical scene.

It may be a lovely thought to beat tarnished swords into environmental plowshares, but proponents of that view are probably due a rude encounter with reality. I would suggest that the Pentagon pork barrel and its defenders in Congress are every bit as entrenched as the Communist bureaucracy in the USSR. Further, there is enough uncertainty over the outcome of events in the USSR to give the defenders of Pentagon pork ammunition for slowing the military rollback.

There is even more uncertainty over the U.S. economic future. A huge social agenda is splayed before the American public today that covers everything from drugs, crime, and education to AIDS and homelessness. Yet the United States is grappling with some of the worst economic problems in its history—trillion-dollar debt, unshakeable budget and trade deficits—amid these compet-ing concerns. Sen. Al Gore (D.-Tenn.) may have a sexy agenda for his run at the Presidency in 1992 with his proposal for a Strategic Environmental Initiative, but a recession could bring that high-flown notion to ground very abruptly.

How much the environmental agenda in Washington commands center stage likely will be determined by the relative economic comfort zone of the country at the time.

A key indicator is what will happen now that President Bush has embraced new taxes, a reversal of his "read my lips" campaign pledge. A careful reading of his comments will show that the President has left himself some rhetorical breathing room. Should a public backlash materialize (are we cynical enough to vote overwhelmingly for a man whose political centerpiece is to not raise taxes when at the same time we think he is lying to us?), or Democrats succeed in turning Bush's reversal to their advantage, Bush is certain to back off and become a born-again tax opponent.

In California, voters have approved a sharp gasoline tax increase. But it would not be wise to read much enthusiasm for new taxes into that. California voters approved that tax hike because it was specifically earmarked for the state's crumbling transportation infrastructure. Raising taxes for cutting deficits or social issues will prove a much tougher sell.

An energy tax, such as one on gasoline or the proposed "carbon" tax targeting fossil fuel consumption for energy security and environmental (global warming) purposes, will prove a litmus test for the industry. If proponents can sell such a tax primarily on its environmental benefits (energy security only seems to be a worry for the average consumer when gasoline prices are spiking), then the industry really has something to worry about. If gasoline or carbon tax proponents fail, then it becomes industry's overriding mission to convince the American public that most initiatives espoused by the environmental lobby are merely taxes in disguise. There is no higher priority for the industry than educating the American public on the costs of Washington's environmental agenda for the average citizen.

Bush's Role

• •

While he was campaigning for President, George Bush promised that the United States would take the lead among the world's nations in tackling environmental problems.

At the same time, Bush vowed to continue the Reagan legacy of reining the growth of government regulation, especially for American business.

That inherent conflict already is being played out in Washington over Bush's first environmental initiatives.

Bush recognized the significance of the environmental renaissance by promoting it as a foreign policy issue at the outset.

With the Soviet military threat eroding, relations with traditional European allies have begun to focus on other issues. It is only natural for the greens, as they consolidate their newly won mainstream political power, to push environmental concerns to the forefront of their agenda again, now that the specter of nuclear war is fading.

Bush took a step in that direction when he proposed that the United States and its allies strengthen ties to the fast-changing governments of eastern Europe and the USSR by offering western technology and assistance on environmental problems.

Further, Bush told National Security Adviser Brent Scowcroft that the environment must be a major consideration in the conduct of national security affairs.

Still, Bush has resisted falling into lockstep with the international community on environmental concerns. He was taken to task by members of Congress who criticized the administration's reluctance to endorse a resolution put forth at a 1989 global warming conference in the Netherlands. The resolution called for specific curbs on emissions of so-called greenhouse gases by the year 2000.

The White House instead would only agree to a compromise declaration calling for greenhouse gas emissions to be stabilized "as soon as possible," thus not tying the United States to specific levels of emissions reductions or within a set time-frame.

Many of the international conferees wanted to freeze those emissions at current levels by 2000.

But the White House, while committing to a need for reducing greenhouse gas emissions, especially carbon dioxide, contends it is too early to commit to specific curbs of greenhouse gases. The administration instead wants to see the results of further study, notably recommendations forthcoming from meetings by the International Panel on Climate Change that began in February 1990. At the opening session of that panel, Bush expressed a need for moving cautiously, stressing the inconclusiveness of global climate change models.

DOE's Mission

• •

Another suggestion of where environmental policy might be headed under Bush is the emphasis it receives in developing U.S. energy policy under the Department of Energy (DOE).

It was no coincidence that Bush's choice for energy secretary, Adm. James Watkins, was driven by the admiral's experience on nuclear issues and hence his suitability for dealing with the scandal over nuclear weapons site cleanup.

The environmental imperative can be seen in Watkins's efforts to develop a National Energy Strategy.

"The National Energy Strategy will serve as a blueprint for energy policy and program decisions," Watkins said. "This strategy will chart our course, set our pace, and provide mileposts by which to evaluate our progress in providing the energy our economy needs, while protecting the nation's health, safety, and environment."

Specifically, that entails what DOE has outlined as the six principles central to its approach on global climate change policy:

• Take aggressive action on those issues on which scientific consensus exists.
• Assess the state of the *science* on issues where *no* scientific consensus exists, and identify areas for further inquiry.
• Where scientific uncertainty exists, move forward with those measures that make sense on other grounds—efficiency, reducing CFCs, reforestation.
• Consider the costs and benefits of any response measures suggested.
• Link responses to scientific and technical information.
• Determine how to evaluate and share technological responses with developing countries.

"The majority of actions that can be taken to reduce greenhouse gas emissions involve the energy sector," DOE said. "Thus it is proper that the Department of Energy play a primary role both in reducing the scientific uncertainty, and in defining the initiatives that can be undertaken—publicly and privately—in response to this threat."

Separately, the NES will focus on "stewardship of the environment."

"Energy production, transportation, and use can be managed in an environmentally safe manner, but recent accidents suggest that insufficient attention is being paid to the requirement of environmental safety. Risks will always be present, but the American people rightly demand that we do a better job of reducing harmful emissions, managing hazardous waste, and improving the technology to prevent and clean up oil and chemical spills."

Whatever the role of environmental values in the NES, the document is likely to reflect a new activist role for DOE, something the industry should welcome. Probably at no other time since its inception has that agency been headed by someone who not only is receptive to what the petroleum industry has to say, but is willing and even enthusiastic about following through with actions.

Still, things move slowly in Washington. The timetable at presstime called for more public comments, reviews, and workshops in summer 1990, senior DOE working groups analyzing and reviewing energy options in September–October 1990, a final submission to President Bush in December 1990, and for Bush to release the final NES in January–March 1991. Even then, the NES will be subject to changes. A new computer model that the initial NES will be based upon won't be finished for two to three years.

Further, no administration is especially activist in the latter half of its term. So many of the positive things industry can look for under the NES might not be implemented until well into a second Bush term, if it occurs.

Justice Gets Tough

. . . .

The environment will have a higher priority at the Department of Justice in the 1990s as well.

In 1989, Donald Alan Carr, acting assistant attorney general for the Department of Land and Natural Resources, warned the nation's business community:

"Do not ever deliberately violate the nation's environmental laws. If you do, you will be at risk not only of serious financial penalty, but also, for the individual decision-makers involved, it is highly likely that there will be some hard jail time."

Since the Department of Justice inaugurated its environmental crimes division in 1983 until midyear 1989, it netted more than 540 indictments, 412 pleas/convictions, $23 million in fines, and 254 years in jail terms.

Justice's prosecution of environmental crimes accelerated sharply with the advent of the Bush regime. In the first six months of 1989, the environmental crimes division garnered 72 indictments, 87 pleas/convictions, almost $10 million in fines, and 36 years in jail terms. Compare that with 1988, the last year of the Reagan presidency, when Justice obtained 124 indictments, 63 pleas/convictions, $7 million in fines, and 39 years in jail terms *for the full year*. Note the higher ratio of convictions to indictments, stiffer fines, and longer jail terms under Bush versus Reagan.

That degree of prosecutorial zeal is certain to continue accelerating, especially as the Clean Air Act (CAA) revisions become law. Under current law, CAA violations carry misdemeanor penalties. It is likely that the EPA will seek, and Congress will approve, a three-year felony for CAA violations, with a provision for as much as 15 years jail time for reckless disregard for safety and health. Such tougher penalties for environmental crimes were recently added to Superfund and Clean Water Act laws.

Further, more stringent new federal sentencing guidelines, stepped-up enforcement by new environmental crime units at EPA, Justice, and the Federal Bureau of Investigation, and efforts among states' attorney generals and local district attorneys to coordinate investigations of environmental crimes, all add up to a nationwide push to use the law to bring environmental transgressors to heel in a way that echoes the Bush administration's war on drugs.

Justice claimed it obtained guilty pleas or convictions in 107 environmental criminal cases in fiscal 1989, a record.

That was a 70% jump from the previous fiscal year, even though the number of indictments actually declined by about 20%, reflecting a high level of pending cases. The new prosecutorial zeal may have begun under Reagan, but it certainly will accelerate under Bush.

Congressional Pressures

• •

On the Congressional side, giving the environment a higher profile in one's campaign portfolio is not just a plus, it may prove essential to political survival. Witness the stampede among representatives and senators—more than 275 at last count—to participate in the Congress' largest caucus, the environmental and energy study conference. The caucus will focus on such issues as global warming, acid rain, and alternate fuels.

The environmentalist groups will continue to hold politicians' feet to the fire on environmental issues. One way is the approach undertaken by the Public Interest Research Group, a Washington advocacy group supporting the environmental lobby. PIRG issued press releases in fall 1989 after a series of House committee votes on CAA revisions listing the campaign contributions from various business political action committees.

Naturally, those representatives receiving the largest contributions from the business PACs were those linked with amendments supported by business— notably a key vote to scuttle a mandated number of alternate fueled vehicles that was actively sought by automakers and the oil industry. This sort of tactic is nothing new, but it is likely to become more commonplace as senators and representatives increasingly are put to a sort of litmus test of environmental purity (or impurity)—with the unspoken assumption that some politicians are lining their pockets to look the other way as polluters are allowed free rein to despoil the environment. Of course, PIRG never listed the amount of contributions the sponsors of tougher air emissions laws received from the environmental lobby's PACs. That would be tantamount to acknowledging that the environmental lobby is, after all, a special interest.

The Clean Air Debate

. . . .

A good litmus test for gauging the direction of where federal environmental policy will head in the 1990s is the debate over air quality issues that is being forged as this book goes to press.

Air quality will likely be the area of federal policy that will have the most far reaching and pervasive effect on the petroleum industry in the 1990s.

President Bush's proposed CAA revisions in 1989 sparked the beginning of heated debate that will mark the Washington scene throughout the 1990s.

Bush's CAA proposals initially featured a mix of market incentives spotted with the usual heavy-handed command and control imperatives that are the lifeblood of environmentalists and their proxies in Congress and the federal bureaucracy (Fig. 4–1).

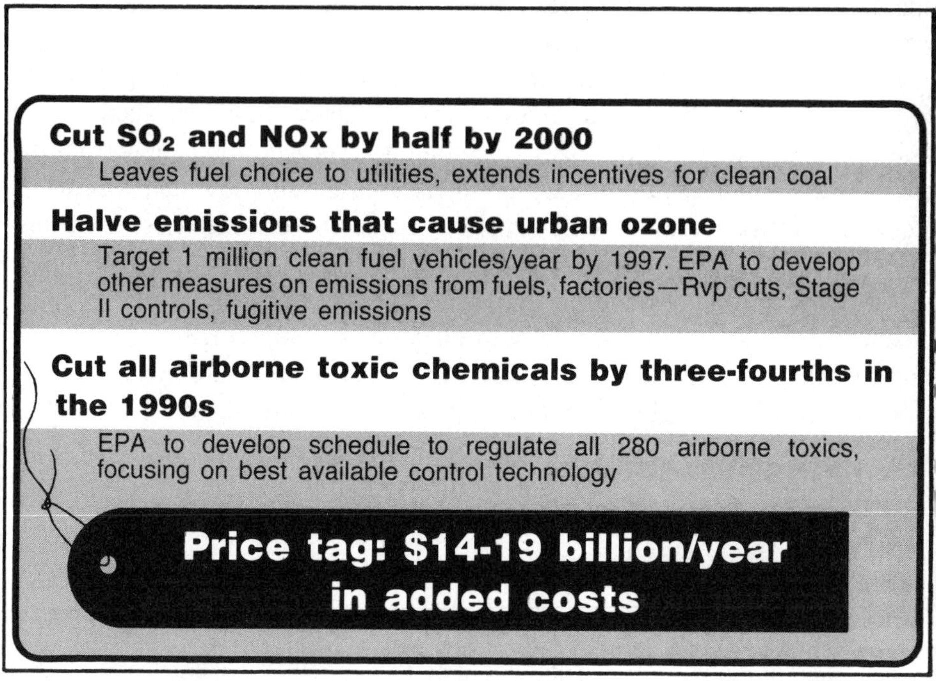

Cut SO$_2$ and NOx by half by 2000

Leaves fuel choice to utilities, extends incentives for clean coal

Halve emissions that cause urban ozone

Target 1 million clean fuel vehicles/year by 1997. EPA to develop other measures on emissions from fuels, factories—Rvp cuts, Stage II controls, fugitive emissions

Cut all airborne toxic chemicals by three-fourths in the 1990s

EPA to develop schedule to regulate all 280 airborne toxics, focusing on best available control technology

Price tag: $14-19 billion/year in added costs

Figure 4–1 Bush's clean air goals, proposals.

Bush and EPA Administrator Reilly described the administration bill as one that would provide market approaches to controlling air pollution.

However, the White House approach also called for the government to mandate the adoption of alternate fuels and vehicles. It would have required automakers to phase in large numbers of alternate fueled vehicles in those areas that have not met EPA's national ambient air quality standards (NAAQS) for ozone, carbon monoxide, nitrogen oxides, sulfur dioxide, lead, and particulates. That would begin in 1995 in the urban areas with the worst carbon monoxide problems.

At the same time, refiner/marketers or other fuels producers would be obliged to phase in production of "clean" alternate fuels. The measure defined clean alternate fuels as methanol, ethanol, natural gas, propane, electricity, or other motor vehicle fuel or power sources such as reformulated gasoline if it has comparably low emissions as determined by EPA.

Target areas are Los Angeles, Houston, New York, Milwaukee, Baltimore, Philadelphia, Greater Connecticut, San Diego, and Chicago.

It would have required EPA, within 18 months of enactment, to issue rules requiring 500,000 clean fuel vehicles to be sold in each area beginning in model year 1995. In 1996, that would increase to 750,000 clean fuel vehicles, and in 1997–2004, 1 million vehicles/yr.

In addition, the original version of the Bush CAA bill would call for gasoline sold in the areas with the worst carbon monoxide levels to contain sufficient oxygen to meet the NAAQS for carbon monoxide. EPA would determine that the required alternate fuel would be made otherwise available and would require that all service stations selling an average of 50,000 gal./month of motor fuel also offer at least one clean alternate fuel. The agency would also have the authority to require sales of those fuels outside the targeted urban areas. If those urban areas could cut air emissions to meet their required NAAQS levels through other measures, then they would be allowed to withdraw from the alternate fuel vehicle programs. Correspondingly, nontargeted urban areas could enter the clean fuels program if a state requested it. In addition, a state could ask EPA to increase the number of required clean fuel vehicles if that would enable the area to meet its targeted NAAQS.

The market incentive aspect of the Bush air quality program involves an EPA rule allowing automakers to trade emissions and fuel producers to pool fuels, thus enabling both industries the flexibility to meet emissions standards efficiently.

EPA would be able to trim or postpone the clean fuel vehicle program by as much as two years, if the agency found that would either substantially advance the technology, improve benefits, or lower costs—or if appropriate because of a likely recession.

EPA also would have sweeping authority to regulate vehicle emissions—setting carbon monoxide emissions at 10 g/mi at 20°F for new cars and trucks, cutting tailpipe emissions of nitrogen oxides to 0.7 g/mi from 1 g/mi, requiring vehicle emission control diagnostic systems to alert for repairs, setting gasoline volatility at 9 PSI Reid vapor pressure or less as needed in certain areas, and reducing sulfur dioxide levels in diesel fuels.

The White House CAA bill also called for devising a timetable and methodology for regulating air emissions of toxic substances from stationary sources.

Specifically, that calls for maximum achievable control technology (MACT), or the best emissions control achieved in practice by a comparable source. MACT would be applied at first to the air toxics thought to pose the greatest risk to health. Seven years after MACT was in place, EPA would review sources to determine if further controls were required. States or regions would be required to provide EPA with comprehensive programs for permitting stationary sources of air toxics emissions within three years of enactment.

The third major area of air quality concern addressed by the Bush proposals covers acid rain. It would require electric utilities to cut sulfur dioxide emission levels by 9 million tons/yr and non-utility sources to cut them by 1 million tons/yr by the year 2000, from levels in the 1980s.

Electric utilities burning fossil fuels that generate at least 100,000 kw after 1995 must limit sulfur dioxide emissions to 2.5 lb/million BTUs. Smaller plants receiving emissions allowances under that standard would be allowed to trade them within a state or utility system.

Under a second phase, power plants larger than 75,000 kw generating capacity emitting more than 1.2 lb/million BTU would receive permits equal to 1.2 lb/million BTU times their average annual fuel consumption from 1985 to 1987.

Plants below that emission level would be allowed to increase operating capacity but only if they maintained the low emission standard.

Those seeking to construct new power generating capacity would be required to trade for offsetting emissions allowances by shutting down plants or using surplus emissions credits. However, plants that employ clean coal technology will earn emissions allowances to permit expansions. Clean coal plants would also win an extension of the second phase deadlines by three years to 2003, among other incentives.

U.S. petroleum industry associations as well as other business lobbying groups were not enthusiastic about Bush's first version of his proposed CAA revisions. They mainly expressed concern about the sweeping powers granted to EPA. Perhaps the most vocal objection came from the American Petroleum Institute, claiming that it was common knowledge that Bush and EPA's Reilly favored methanol as the "clean fuel" of choice.

API contended that the Bush proposals would force motorists to use methanol, which it described as "an extremely expensive and inconvenient alternate fuel that involves serious unresolved questions about its environmental, safety, health, and energy security implications."

Such a mandated fuel also would run counter to Bush's vow to let the marketplace determine which fuels or technologies will be used to achieve air quality goals, API said.

Oil and auto companies lobbied against the clean fuel vehicles portion of the Bush proposal, winning a House subcommittee vote on an amendment to weaken that provision. It dropped the requirement calling for 1 million clean fuel vehicles per year during 1997–2004, instead merely allowing automakers simply to certify that they were capable of building the alternate fueled vehicles.

Gas industry associations lauded portions of the proposal, noting the opportunity for gas market share growth through increased utility consumption of gas directly or cofired with coal to cut acid rain-related emissions and through increased use of compressed natural gas as an automotive fuel.

The environmental lobby attacked the Bush proposals to provide market incentives for cleaning up air emissions. They were aghast at the concept of applying cost-benefit measurements to any air quality measures.

Another early casualty of the President's CAA proposals was the vehicle emissions trading program. That proposal called for allowing automakers to trade emissions reductions across vehicle categories and among manufacturers. If a carmaker produced a vehicle that polluted less than the new federally mandated level, it could sell that "credit" or use it up in the production of another vehicle. That would even provide additional incentives to try alternate fuels.

The American business community largely supports such a market incentive approach, mainly in view of its economic benefits. For example, assume that a law requires two businesses each must reduce emissions by X amount. For Business A, it costs $10X to meet that standard. Business B, however, must invest $1000 to meet that standard. Without market flexibility, the cost to the economy to reduce emissions by X becomes $1010X. But if emissions offsets were available to buy at $100X, Business B would find it cheaper to buy offsets than cut emissions. But Business A would find it 10 times cheaper to cut pollution. That would result in the same or greater environmental benefit at a lower cost to the economy.

The debate over command and control approaches vs. market incentive approaches heated to boiling in 1990 as the prospect for passage of some version of a CAA bill became inevitable before Congress adjourned in fall 1990.

Although opposing groups' level of discourse on Clean Air is rancorous, Congress, the White House, and the environmental lobby all want the CAA reauthorized now.

The final bill, approved by Congress in October 1990, bore the hallmarks of compromise between the two approaches enough to gain passage but also enough to keep very much alive the rancor among opposing factions.

The question of cost is at the heart of differences between the White House's approach and Congress's. A bill marked early in 1990 by the Senate environment committee pointed to the wide variance in costs between the two approaches—even to a wide variance in assumptions over how to calculate those costs.

Bush maintains the Senate bill would cost American industry $40 billion/yr by the end of the century, compared with his own bill's estimated price tag of $19 billion/yr to achieve the same cut in air pollution.

The President earlier indicated he is likely to veto any bill that will cost appreciably more than his own, create new taxes, or not allow for reducing emissions through a least cost-per-ton of pollutant approach. If his personal popularity is sustained through 1991 at the level he enjoyed at the beginning of his term, he is almost certain to be sustained by Congress in that veto.

The Senate environment committee bill avoided requiring alternate fuel vehicles in urban areas not in attainment with federal ozone standards, as the Bush bill in its original form did. The committee left that up to the Senate at large to be included as a later amendment, notably because it will be the most controversial aspect of the bill. Bush himself indicated that the White House may attempt to reinsert that provision in his bill.

It is likely that the fate of the administration's zeal for mandating that a big chunk of the nation's vehicle fleet runs on methanol and the like rests on the outcome of the Persian Gulf crisis. If it is not resolved, and gasoline prices remain high, public clamor probably will oblige a sizable alternate fuel vehicles program, and Bush will have little choice but to climb back on the bandwagon. If oil and gasoline prices fall again, and there are no severe pollution episodes in early 1991, the oil and auto industries will probably have dodged a large caliber bullet. It will be surprising if there is not at least some modest version of an alternate fuel vehicles program in the final CAA reauthorization, still waiting for Bush's signature as this book went to press. A safe wager might be on a more-than-token but not-too-hefty demonstration program that will show the administration's and Congress's commitment to alternate fuels without running afoul of unemployment and budget concerns as the U.S. economy flutters between stagflation and recession.

The Senate bill also called for:

- Compliance in 100 nonattainment areas for ozone, 40 areas for carbon monoxide, and 58 for particulates within 5–20 years, depending on type and severity of pollution.
- Fees of $75/ton of air emissions, with fees to be collected and used by local air pollution control agencies.

- Tougher tailpipe emission standards for new cars and trucks, halving the permissible level of emissions by 2003.
- Stricter fuel economy standards.
- Service stations in the worst urban ozone areas to install Stage II volatile organic compound (VOC) emission recovery equipment (such equipment in the form of vapor recovery nozzles on gasoline pumps have been widely derided, hated, and eventually accepted as a disagreeable fact of life for years in California).
- EPA to control emissions of 200 toxic chemicals from industrial plants and tighten safety standards for accidental release of about 50 very hazardous substances.
- Utilities to trim sulfur dioxide emissions to 10 million tons/yr by 2000, where they will remain capped, thereby obliging new plants to obtain emissions offsets before increasing capacity.

The industry's biggest concern under the Senate Clean Air bill, however, is an amendment tacked on by Sen. Tom Daschle (D.-S.D.) that sets standards for reformulated gasoline and requires its use in nine cities that did not meet ozone attainment standards. It will phase in during 1992–94.

The real problem with the Daschle amendment is that it specifies a formula for reformulated gasoline: limits of 25% for hydrocarbon content, 1% for benzene content, and a minimum of 2.7% oxygen content by weight. The last is especially nettlesome because the only way industry is likely to meet that oxygen standard is by adding ethanol. Refiner-marketers fear the subsidized additive's lack of acceptance by the public and the loss of gasoline market share, not to mention the prospect of increasing certain unwanted emissions because of ethanol's increased volatility. Perhaps the biggest concern is the prospect of adding as much as 25¢ to the cost of a gallon of gasoline, and what that will do to overall demand.

The House has its own CAA reauthorization bill under way that has similar goals, if slightly different approaches, compared with the Senate bill. The major point of interest is that the House bill offers industry a bit more flexibility in meeting gasoline requirements.

The House bill required:

- Reformulated gasoline to cut emissions 15% by 1995 and 25% by 2000.
- Minimum oxygen content of 2.7% by weight in winter months in 44 cities exceeding carbon monoxide standards.

- Lower nitrogen oxide emissions, to 0.4 g/gal from 1 g/gal by 1996 and half that by 2003 if EPA determines it necessary.
- Cuts in hydrocarbon emissions to 0.25 g/gal from 0.41 g/gal by 1996, and half again in 2003 if needed.
- Sponsoring a pilot program in California beginning in 1994 that would call for 750,000 alternate fueled cars by 1997.
- Best available control technology by 2000 for emissions of 200 toxic chemicals.
- A $250 million unemployment compensation package for coal miners and plant workers who lose their jobs because of the new rules.

That last item could prove the CAA reauthorization bill's undoing, if it survives the House-Senate conference committee efforts at a compromise bill in summer 1990. President Bush has promised to veto any bill containing such a relief package.

Another early problem for the House bill was a subcommittee effort to obtain a cost-sharing provision for midwestern states heavily reliant on burning coal in their utilities. In these days of $3 trillion budgets, another such obviously provincial subsidy is not likely to survive.

That brings to bear a crucial point, one the petroleum industry must not only treat as gospel, but evangelize that scripture to the American public with the same fervor and dedication that the Neodruids have had in spreading their dogma:

What is all of this going to cost the American consumer?

It is critical for the petroleum industry to convey to the American public the costs of extreme environmental regulation in the most concrete terms: jobs lost, mounting trade deficit, reduced international competitiveness, increased product costs, and higher taxes.

It won't suffice to complain about the petroleum industry's increased costs in general terms. The industry will forever be linked in the public's mind with the "obscene profits" that it was accused of raking in in the wake of the oil price spikes of the 1970s. A quick look at the price spikes for gasoline after the Exxon Valdez spill and for heating fuels in December 1989's cold snap will tell you that that sentiment is still very much alive.

If the U.S. petroleum industry cannot successfully get the message through to the public (not just the politicians) that the environmental agenda will mean big out-of-pocket expenses for John Q. Public individually, then its own economic survival may be in question.

While Bush put a price tag on the Senate CAA bill of $40 billion/yr, Clean Air Working Group Administrator Bill Fay questions whether Bush's own estimated CAA tab is valid.

"We need a full and careful analysis of how much the bill costs," Fay said in an August 1989 press release. "The original $14–$19 billion range is no longer valid. In fact, in his June 23rd newsletter, economist Michael Evans noted the annual price tag could approach the $60 billion range."

"If this figure is added to the $33 billion already spent each year on clean air, the resulting $93 billion cost would be greater than the annual budgets of any federal departments except Defense, Health and Human Services, and Treasury."

The Office of Technology Assessment, noting that about two-thirds of the reductions in VOCs needed to meet the ozone standard in all areas can be achieved using available controls, estimates the price tag for this effort at $4.4–$7.8 billion/yr in the mid-1990s and $8.8–$13 billion/yr a decade later.

Business Roundtable put the Senate bill's cost to industry at $54 billion/yr: $25 billion to cut air toxics, $24 billion for ozone reduction, and $5 billion for acid rain compliance.

That total could be as high as $104 billion and force many plants to close or restrict production, said Business Roundtable. Further, if the most stringent air toxics rules are put into place, their costs alone could reach $55 billion/yr.

Advocates of a strengthened CAA measure counter by pointing out the American Lung Association's claim that air pollution costs Americans $40 billion/yr for health care.

Is there a Hobson's choice here? One key task for the petroleum industry in seeking to deal with these issues is to sort out the variables in the data and perhaps fund an independent (such as the National Academy of Sciences) study of environmental costs and benefits.

It ought to suggest something to petroleum companies as to why the environmental community is so adamantly opposed to assessing environmental measures on a cost-benefit basis. The environmental lobby, while decrying that very approach, is itself dredging up statistics in favor of its own arguments, in essence telling John Q. Public that these environmental benefits would reduce some of his out-of-pocket expenses.

That game is being played with special alacrity in states where environmental activism is on the rise.

State Government Activism

· ·

Activism on environmental issues is accelerating among state governments. Since the Exxon Valdez oil spill, Alaska has taken steps on a host of environmental issues generally revolving around the transportation of oil via TAPS and tankers that will cost the petroleum industry hundreds of millions of dollars in the 1990s. Those measures will be covered in a later chapter on transportation issues.

Some of the activism is merely pork barrel politics in disguise. For example, Nebraska lawmakers passed a bill in spring 1990 that requires that all reformulated gasoline sold in the state contain at least 3.1% oxygen by weight. That may make sense in areas with severe carbon monoxide or ozone problems, but Nebraska is not one of those areas. The rationale? Mandating a 3.1% standard for gasoline essentially means a mandate for heavily subsidized ethanol—made from Nebraskan corn, of course.

Even states typically friendly to the petroleum industry aren't shying away from activism on environmental issues. Texas is mulling new oil spill legislation in the wake of the June 8, 1990, explosion and fire that ravaged the Mega Borg tanker off Galveston, Texas. Louisiana in 1990 floated more than 150 environmental bills that include a 5¢/bbl tax on oil for a $10 million oil spill liability fund, a 20¢/bbl fee on nonhazardous oil field wastes, and linking industrial tax exemptions to environmental behavior.

Some states are acting in combine to effect environmental measures. Rhode Island, Connecticut, New York, and Massachusetts are cooperating on regional oil spill protection and liability issues. Delaware, Maryland, New Jersey, North Carolina, Pennsylvania, Virginia, and the District of Columbia created an association to address air quality issues, notably gasoline volatility. A similar association was formed in 1989 by New York, Pennsylvania, Connecticut, Rhode Island, New Jersey, Massachusetts, New Hampshire, and Vermont.

States eyeing new oil spill legislation include Delaware, New Jersey, Connecticut, Rhode Island, New York, Maryland, Alabama, South Carolina, and Florida.

Florida also extended a ban on leasing off its coast, and its lobbying led industry to agree to a tanker-free zone off the Florida Keys.

Efforts are afoot to mandate conversion of state vehicle fleets to alternate fuels in Louisiana, Texas, Tennessee, and Kentucky.

Iowa and Tennessee are implementing new strictures on aboveground and underground petroleum products storage.

These, of course, are only the beginning. States are likely to become more activist on environmental issues, even when the motive is not environmentalism. Alaska used the window of opportunity immediately after the Exxon Valdez spill to push through a reformulation of oil and gas severance taxes that already cost industry several hundred million dollars in added taxes in 1989 alone. Although Alaskan legislators have been hungrily eyeing more taxes on oil for years—and especially since oil prices began declining, slicing the revenues of a state almost 90% dependent on oil money for its revenues—industry had always been able to turn the tide—until the Exxon Valdez.

So the petroleum industry has another powerful incentive to be religiously scrupulous in its environmental track record: polluting tends to draw public ire and therefore political opportunities to bleed more money from corporate coffers.

Activism on environmental issues is a relatively new thing in some states. But there is one state where environmental activism long has been an article of political faith. And in California, that has ascended to the status of state religion.

California: Blueprint for the Future

• •

In soliciting comments for its 1989 biennial fuels report, the California Energy Commission took an environmental slant, contending that production and consumption of energy are "generally hostile to the environment."

"Yet energy is an essential component of economic activity. Historically, the costs of preserving the environment have not been included as a cost of energy use. Now, increasing public concern about the societal impacts of pollution have focused government and industry attention on the connection between energy and the environment."

The agency then solicited testimony on policy strategies to limit, control, mitigate, or remedy adverse effects of energy production, conversion, and transportation on California's environmental quality. Much of the subsequent focus was on alternate fuels, conservation, and even environmental concerns outside the state—such as the environmental repercussions of meeting the state's energy needs with imports from other regions, namely exploring for and producing Alaskan oil that is shipped to California and burning fossil fuels in other states to produce electricity that is imported into California.

The CEC estimates that oil's share of California's overall primary energy demand will slip to 50% by 2007 from 54% in 1987—of which three-fourths went to transportation. However, the absolute volume of oil demand in the state will continue to rise with population growth.

About half of California's oil supplies come from within the state, the rest mostly from Alaska, with a small volume of foreign oil imports. However, California's oil production is declining again, and Alaska's North Slope has entered into a steep decline with almost no chance of being reversed this century. That would leave California faced with the prospect of becoming increasingly vulnerable to a dangerous dependence on foreign oil imports.

It bears mentioning that Offshore California has the biggest untapped, *identified* conventional oil resource in the United States—perhaps in the world, if you factor in need and prospects for access.

It also bears mentioning that California ranks behind only the United States and the USSR in consumption of gasoline—a seemingly unslakeable thirst at more than 1 billion gal/month.

The CEC, which is charged with ensuring a reliable and economic energy supply, also noted that energy supply and consumption are critical to maintaining a competitive state economy—one that, incidentally, stands as the sixth-largest in the world.

One would think, then, that the agency in California dedicated to ensuring a reliable and economic energy supply, would want to promote indigenous energy resources, correct?

Except for a fleeting reference to its 1987 biennial fuels report, which recommended state funding and support for enhanced oil recovery research, there is no reference in CEC's biennial fuels report for 1989 to encouraging the exploration and development of California's vast oil resources.

Instead, the CEC promotes conservation and the use of alternate fuels to back out oil. The economic rationale is put forth that the cost of dependence on oil will rise as oil prices rise in the 1990s owing to tightening supply/demand factors and associated higher environmental mitigation costs.

This stance apparently disregards the costs of conservation. There is much data to support the contention that virtually all of the relatively low-cost conservation measures have been taken. There are other costs associated with lost economic competitiveness, lost jobs, and higher health care and insurance costs (the "sick building syndrome" seen with excessive insulation/conservation measures, and the higher fatality and severe injury rate with smaller cars).

It is also obliquely presumed that alternate fuels costs will somehow be less than those of conventional fuels with added environmental constraints, not to mention the possibility of new environmental hazards, such as those posed by the CEC's preferred alternate fuel, methanol.

It is no secret that the CEC has for several years been touting methanol as a panacea for the state's energy dependence and air quality concerns. It has implemented a program to demonstrate a methanol-based fuel in fleet vehicles in northern and southern California, with ARCO, Chevron, Exxon, Mobil, and Shell participating by providing the methanol mix at service station pumps, and the state buying methanol-fueled test vehicles for the demonstration program.

In other words, for the CEC, oil use itself is the problem, and conservation and methanol are the solutions.

That attitude already is seeping into the federal government. The U.S. DOE in November 1989 adopted new national refrigerator energy conservation standards that were patterned after standards established by the CEC in 1984. In addition to reducing emissions of CFCs, the new standards will curtail emissions created by burning fuel to produce electricity. DOE estimates that by 2015, the new refrigerator standards will cut almost 10 million tons of carbon dioxide emissions, 126,000 tons of sulfur dioxide, and 95,000 tons of nitrogen oxides through reduced demand for electrical power.

It is in the area of air quality that California's environmental policies are the most stringent, and most likely to provide a blueprint for the rest of the nation.

The South Coast Air Quality Management District (Scaqmd) has adopted a new antipollution program for cleaning up air quality in the South Coast basin—which covers 13,350 sq mi in Los Angeles, Orange, and parts of Riverside and San Bernardino counties—that is frightening in its sweep and scope.

The 5,000-page plan calling for more than 160 specific measures seeks to attain compliance for various air emissions within state and federal guidelines by 2007.

Aside from the usual control measures on various businesses and industries, there are strikingly invasive measures targeted at the average citizen. If Scaqmd has its way, Angelenos will be forbidden to use gasoline-powered lawnmowers, have backyard barbecues, use drive-through windows at fast food restaurants, use aerosol hair sprays and deodorants, or buy bias ply tires.

But the real thrust of Scaqmd's Air Quality Management Plan (AQMP) is an effort to enforce a massive switch to alternate, "clean" fuels in the 1990s and beyond. It locks into a specific policy what has long been the intent of California's air pollution control agencies, as well as many local governments: the ultimate elimination of the petroleum industry from southern California. This is not hyperbole; it is a charter goal of the local/municipal governments association, Southern California Area Governments (SCAG), and one that SCAG has espoused in official proclamations.

Furthermore, it promotes a program calling for development of major technological breakthroughs to meet federal and state air quality standards. It states: "If sufficient technologies to achieve the standards are not identifiable by the mid-nineties, a contingency plan will be developed for replacing high-polluting industries with low-polluting industries having equivalent employment potential."

Quite a grand design, isn't it? George Orwell ought to be spinning in his grave.

The AQMP calls for 40% of California's vehicles and 70% of the state's trucks to be capable of running on methanol or other "clean fuels" within 10 years. Within 20 years, all vehicles will run on electric power or an equally low-polluting power source.

The AQMP's exorbitant costs for control measures alone may accomplish the job of driving out the petroleum industry even if the forced switch to alternate fuels doesn't. The AQMP is expected to cost the petroleum industry more than $330 million the first year alone.

Under the AQMP, municipal bus fleets would have to switch to clean fuel vehicles by 1991. Although the AQMP does not specify methanol as the required clean fuel, all of the plan's specifications are based on methanol. By 1993, that switch would extend to taxis, rental cars, and other commercial fleets. By 2009, the average Angeleno will find it virtually impossible to buy a new car that is not designed with emission levels equal to a methanol fuel's level.

Ford Motor Co. has already delivered the first units of its flexible fuel Crown Victoria to the CEC for a demonstration program, as has General Motors its new variable fuel Chevy Corsica. Both are able to run on methanol, gasoline, or a mix. The two automakers expect the models to be commercially available by 1993.

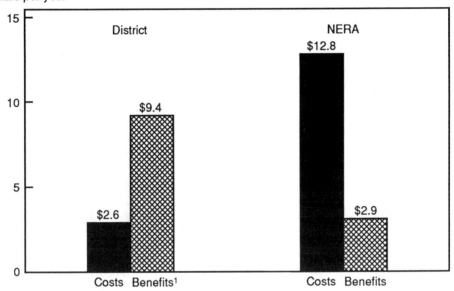

¹ Benefit estimates are from Hall et al., 1989.

Source: National Economic Research Associates Inc.

Figure 4–2 AQMP's annual costs and benefits.

Scaqmd contends the AQMP will cost almost $4 billion/yr. But a study by National Economic Research Associates Inc. put the cost at closer to $13 billion/yr. That works out to about $2,200/yr added to the cost of goods and services for each household in southern California (Fig. 4–2). And most of that burden would fall disproportionately on the lower income groups, the NERA study found (Fig. 4–3).

Even less certain are the ultimate costs to southern Californians of Orwellian mandates such as requiring ridesharing, staggering work hours, and constructing housing closer to work sites to cut commuting.

Then there are the true unknowns covered by the third phase of the three-state program, which calls for implementation of technologies that have not been developed yet.

The debate over costs will continue to rage. Los Angeles County Supervisor Mike Antonovich has claimed that the program is not cost-effective and could

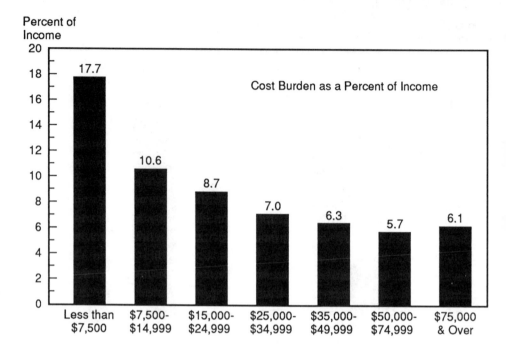

Source: NERA

Figure 4-3 The draft AQMP is like a sales tax that hits the poor hardest.

cause the loss of as many as 150,000 jobs. Scaqmd figures, on the other hand, that the plan would create 80,000 jobs, primarily for minority workers in transportation. That last one is a puzzler, since the mass transit system in southern California would have to undergo a massive expansion just to be deemed adequate.

Scaqmd also has come back with a computer model and study that purported to show that adopting the AQMP would result in a health cost savings of $9.4 billion/yr. But NERA's study concluded that the actual health benefits from implementing the AQMP would be only about $1.6–$6.7 billion/yr.

The Western States Petroleum Association (WSPA) developed its own plan for reducing ozone in southern California within 10 years at a much lower cost and in a much less disruptive way. The association also noted that research is under way by Detroit to develop a cleaner auto engine system that could meet the AQMP's tailpipe emissions standards by the turn of the century.

Scaqmd rejected the WSPA proposal as inadequate for meeting federal standards for other emissions such as nitrogen oxides and particulates.

Meanwhile, the boldness of the Scaqmd thrust is encouraging California's environmental community to broaden its power base. In the San Joaquin Valley, the Sierra Club and other groups are stumping for a plan put forth in the state senate to form a valley-wide air quality management district akin to Scaqmd, consolidating all air emissions permitting into one regional district and thus wresting control from the local level. It should give anyone in the petroleum industry pause to consider another Scaqmd riding herd on Kern County, which, if it were a state, would rank only after Alaska, Texas, and Louisiana in terms of oil production. In short, there may be more AQMPs in the industry's future in California.

Meanwhile, Scaqmd and SCAG have adopted the AQMP, and California's Air Resources Board (CARB) has approved the plan with essentially no qualifications. The AQMP has now gone before EPA for approval and thus the green light on full implementation. Although a Bush administration agency might ordinarily be expected to gag on something as totalitarian in nature as the AQMP, EPA may not have much choice. Under the CAA guidelines, EPA was obliged to step in with marching orders of its own if California did not develop a plan to achieve compliance within a certain time. The state already is under a construction ban related to new source emissions because it ran afoul of EPA deadlines for compliance. Although it has been granted dispensations in the past, the state recently stood the chance of losing federal matching highway funds for its inability to clean the air. EPA's hands may be tied on the matter.

EPA gave an early indication of how it views the AQMP, however:

> Attainment of the ozone NAAQS appears to require, in the case of the South Coast basin, extreme restrictions on the use of fossil fuels and reactive hydrocarbons.

> As a practical matter, immediate attainment is impossible, and a five-year plan would impose requirements so draconian as to remake life in the South Coast basin, prohibiting most traffic, shutting down major business activity, curtailing the use of important consumer goods, and dramatically restricting all aspects of social and economic life.

Meanwhile, CARB is hardly standing by idly. The agency wants to cut levels of sulfur and aromatics in diesel fuel in the early 1990s.

It would extend Scaqmd's existing diesel sulfur limit of 0.05% by weight throughout the state effective in 1993. Currently, diesel sulfur in California averages about 0.28% by weight.

CARB also wants to rein aromatics levels in diesel to 10% by volume, unless refiners can develop an alternative diesel fuel that could match emissions reductions under the 10% rule.

EPA's own proposal to cut diesel sulfur to 0.05% and aromatics to 20% would not take effect until the mid-1990s.

CARB also wants to reduce the aromatics content of gasoline, especially benzene, to about half of its current level of 1.6% by volume.

With all this regulatory action afoot, one would think California's environmental lobby would be relatively content to let some of the political dust settle a bit before launching another salvo.

Just as totalitarian in concept was the new Environmental Protection Initiative sponsored by California Attorney General John Van de Kamp and Assemblyman Tom Hayden. It was defeated by a two to one margin in November 1990, the victim of a suddenly sagging economy and widespread dislike for Hayden. But elements of it are likely to resurface.

The EPI's elements were largely aimed at petroleum and chemicals industries, but also targeted pesticide use and public sewage treatment. Much of it was redundant, echoing elements of Prop 65 and the AQMP.

What really chilled about the EPI was that it provided for election of an Environmental Advocate, a sort of Green Czar. The EA would conduct investigations and studies and bring legal action to ensure compliance with the state's environmental laws. The EA would also recommend policies to ensure environmental protection as well as administer grants to develop alternatives to pesticides and chemicals that are phased out by the initiative.

The EPI also would have required the state to cut emissions of carbon dioxide 20% by 2000 and 40% by 2010 under its plan for reducing greenhouse gases. The initiative did not spell out how this would be accomplished, but probably would have relied on the AQMP. Also as part of its global warming plan, the EPI called for a $300-million bond program to acquire redwood forests and for extensive reforestation in urban and rural areas. It further would have banned clear-cutting of old growth redwoods and required developers to plant one tree for every 500 square feet of new development in the state.

Further, the EPI would have phased out by 1997 use of CFCs unless necessary for medical reasons. CARB would be able to grant limited extensions of CFC use for as long as two years.

The measure also would have resurrected an oil spill prevention and cleanup plan largely along the lines of one introduced in the state legislature by two California Coastal Commission members (one of them Van de Kamp).

There is likely to emerge a separate oil spill bill from a compromise of separate bills in the senate, and the assembly that would assess a 25¢/bbl fee on all oil transported or produced off California's coast, exempting most independents. It would build an emergency oil spill response fund of $150 million.

The difference in the EPI's oil spill plan is that it included a provision banning new drilling on leases in state waters except in the event of a national energy emergency (presumably said emergency would have the good grace to signal itself several years in advance to accommodate the lengthy permitting times for offshore drilling and production projects).

Another aspect of the oil spill plan under the EPI would have ordered the State Lands Commission to develop an oil spill prevention plan that includes using tugs to escort tankers, creates emergency stations for disabled tankers, and requires on-site inspection of potential oil spill sources. Further, it would have created a $500-million oil spill prevention and cleanup fund to finance immediate cleanup of a spill, offer loans to spill victims, and restore land damaged by a spill. The fund would be financed through a fee of 25¢/bbl of oil that is transported through waters off California. All companies transporting said oil would be required to have an oil spill cleanup plan that includes the best available containment and cleanup technology. The EPI also would have created an office of oil spill response within the state Department of Fish and Game to direct all spill cleanup efforts.

Getting into Prop 65 territory, the EPI would have set a range of water quality standards and a statewide monitoring program to assess water and sediment quality. In addition, it would have changed existing marine pollutant discharge fees to better reflect the volume and toxicity of a substance that is discharged into state waters. And it would have required industries that routinely discharge into state waters to conduct pollution prevention audits to identify routinely discharged pollutants and establish methods to reduce those discharges.

Much of the EPI was redundant and unnecessary, covered by existing or proposed regulations elsewhere at the state or federal level or within the petroleum industry itself—notably the oil spill program.

There were staggering potential costs associated with the EPI, according to a study by Spectrum Economics Inc., San Francisco, for the California Coordinating Council, a coalition formed by businesses and associations to defeat the initiative. Restrictions on carbon dioxide and CFCs would have raised the cost of and constrained the use of automobiles in California in a variety of ways, including: hiking gasoline prices 25¢–50¢/gal by 2000 and double that amount by 2010; restricting choice of autos to those with fuel efficiencies of 40 mi/gal or better; doubling the price of diesel fuel and possibly resulting in its rendering $90 million or more of installed auto air conditioners useless; and forcing restrictions on auto use if higher fuel prices and increased auto efficiency don't achieve the big cut in CO_2 expected.

What should really chill the petroleum industry is the prospect of a Green Czar who could prove to be a real loose cannon on environmental matters, using the high profile of his "czarship" to fire random volleys at companies or operations in the state without any apparent purview by a legislative body. It was an open secret in California that Hayden was angling for the job himself. Is there any doubt as to whether he would be an activist Green Czar? The EPI would have proven another huge headache for an industry already suffering from terminal environmental migraine in California.

It was something of an understatement when WSPA characterized the initiative as a "political grab bag."

After all, in a press conference called by Jane Fonda in January 1990 to promote the EPI, the actress/activist announced that the disbanded rock group the Go-Gos were reuniting for a benefit concert to raise funds for the initiative.

In La-La Land, you just don't get any stronger validation of a political concept than that.

5

NEW APPROACHES

The petroleum industry must begin to rethink how it deals with environmental concerns.

It is no longer sufficient to merely be in compliance with existing regulations and just observe the letter of the law. What is acceptable today may not be acceptable tomorrow.

Petroleum companies must move beyond compliance into taking the initiative. Taking the initiative goes well beyond mounting a public relations effort. It is not enough to promote warm, fuzzy feelings as with one of those lushly photographed television commercials depicting a rare butterfly preserved by designing a project in a way to preserve its habitat. It is not enough to solicit good will with full-page print ads featuring testimonials by employees frolicking in a nature scene with their kids while they tout the company's concern for the environment.

Those are valid examples of good P.R. But industry won't survive on just good P.R. And it won't survive by just reacting to specific regulations and laws proposed to govern environmental behavior.

Image damage control is one thing, but that will also fall far short of what is needed to grapple with born-again environmentalism. Too often the petroleum industry has resorted to mere public relations bandages to stem a hemorrhage of public disfavor in the case of an environmental, safety, or health problem of any magnitude.

The petroleum industry has a new mandate in facing the public, in regaining a public trust sullied by two decades of getting the blame, directly and indirectly, for many of this nation's economic and environmental woes. All the righteous indignation and setting-the-record-straight and soft-focus cinematography in the world is not going to do the industry one whit of good if the industry is not prepared to change fundamentally the way it does business with regard to environmental concerns. It should then carry that through with specific plans of action, all the while putting forth the most ambitious public education and communications effort any industry has ever seen.

The petroleum industry must take bold action, and it must take that action now!

The public demands nothing less.

To begin with, petroleum companies should undertake a restructuring of their corporate hierarchies to reflect a results-oriented approach to environmental problems; namely, to high-grade environmental/safety/health concerns by creating new posts or elevating existing posts responsible for those areas at the highest levels. That includes the board of directors and senior management, meaning recruiting officers and directors from within and outside whose environmental credentials are unassailable.

More companies will of their own volition add environmentalists to their boards, as Exxon did late in 1989. Exxon named marine scientist John H. Steele as director, largely in response to demands by several large pension funds holding big blocks of Exxon shares. Steele is president and a senior scientist at Woods Hole Oceanographic Institute. One of his tasks is to serve on the Exxon board's public issues committee, formed after the Valdez spill, which will monitor Exxon's environmental policies.

In addition, Exxon appointed one of its top executives as vice-president of environment and safety. Edwin J. Hess, formerly a senior vice-president of international refining/marketing, will oversee Exxon operations worldwide with regard to their effects on the environment. It is the first time that Exxon has installed an environmental officer in such a high-level role.

At the same time, companies ought to be open to the prospect of inviting outside study of its environmental policies. And there should not be a perceived conflict of interest, especially in the event of an environmental mishap. Exxon ran into some controversy a few months after the Valdez spill when it hired former U.S. Attorney General Griffin Bell to conduct an investigation to determine who was responsible for the Valdez spill and to advise its board on what actions are needed as a result of Bell's findings.

Critics blasted Exxon's move as tantamount to employing a hired gun who would just whitewash the investigation. Furthermore, Exxon could claim attorney-client privilege if it wanted to keep some of the investigation's findings out of the public eye. And such moves are seen as attempts to sidetrack more thorough investigations by government. Exxon and Bell denied the claims. And it is a foregone conclusion that the government investigations of the

Valdez spill will be nothing less than exhaustive—including some not-so-friendly inquiries.

The effort also calls for changes in the operating personnel infrastructure of a company, creating new positions responsible for environmental monitoring and problem-solving. That entails recruiting and training a new kind of employee, an environmental specialist—one perhaps deemed antipathetic to an oil company's mission in the past.

It calls for creating new, in-house units and divisions to provide environmental services for the rest of the company.

It calls for rethinking *all* corporate business strategies—not the least of which is a much-needed recognition that there are terrific business opportunities in dealing with environmental concerns. That further argues for diversification into environmental service businesses, just as oil and gas companies diversified into alternate fuels and other natural resources businesses in the 1970s. (That earlier effort proved largely a bad investment with the cyclic nature of other commodities in addition to oil's; however, environmental services will always be in demand—the trick is jumping in early and strongly enough to weather future shakeouts if the field gets glutted with players.)

That last area brings to mind the theme that this book will strive to hammer home for each segment of the industry, what I think is the single most important distinction the industry needs to learn in coping with the Second Greening of America in the 1990s.

An old saw would serve us well here: *There are no problems, only opportunities.* So the key to survival in the 1990s is taking the initiative.

Restructuring

The restructuring of the petroleum industry in the 1990s in response to environmental concerns will not be as dramatic as the restructuring that swept the industry in the 1980s.

That earlier restructuring mainly involved a massive streamlining and consolidation of companies in response to an unexpectedly low oil price outlook.

Restructuring in the 1990s won't be a numbers game, but a redirection of corporate structures to ensure that concern for the environment becomes an integral, inbred, reflexive part of the fabric of day-to-day operations.

Companies are retooling management and operations personnel structures to enable themselves to be more flexible in dealing with environmental problems and more prepared to take the initiative on environmental concerns before solutions are imposed from without.

That change is reflected in changes afoot at API. After an API task force identified the environment as the first and overriding priority for the oil industry in the 1990s, API decided that what was needed was a new tone, a new direction— in the words of Terry Yosie, API's vice-president of health and environmental affairs, "a kinder, gentler API."

With that idea in mind, API then created another task force, chaired by Phillips Petroleum Co. Chairman Pete Silas. This task force will seek out the best ways to develop long-term planning processes and characterize industry attitudes on specific environmental issues.

Once it identifies industry's basic objectives on environmental issues, the association then will determine the best ways to promote them to the public.

As part of that effort, there will be more joint research programs such as the one the petroleum industry has undertaken with automakers on developing cleaner fuels and the engines for those cleaner fuels. Such joint efforts will prove more cost-effective and improve credibility with the public.

There will be more cooperative efforts on environmental concerns outside the petroleum industry, including those with other industries not sharing a common interest such as the auto industry.

A prototype might be the Corporate Environmental, Health and Safety Round-table formed by 10 major U.S. corporations under the guidance of Environmental Resources Management Inc. (ERM). Participants are Allied-Signal Corp., AT&T, Champion International, Ford Motor Co., General Electric Co., Johnson & Johnson, Merck, Occidental Petroleum, Sun Co., and Union Carbide.

This sort of networking on managing environmental issues could be invaluable for petroleum companies, which tend to be somewhat insular (a combination of the industry's legacies of the individualistic wildcatter's spirit and the "good ol' boy" network) in their approach to solving problems.

Inter-industry networking could expose companies to whole new approaches as well as just tips to improve environmental programs. It would expose companies to environmental management philosophies outside the sphere of what would be developed through trade associations, where they might be locked into one approach.

Belonging to a network group could also help a company trim costs through group members finding what was common to environmental programs and thereby avoiding having to start from scratch with each new program.

The ERM Roundtable also created a directory of management tools developed among the member companies. A company with a specific problem could thus review what management tools other companies had employed to deal with a comparable problem. That could even extend to buying rights to use one of those tools or putting together a consortium with other group members wanting to use that tool and develop a subgroup to adapt that tool to their special needs.

An example of a pooled management tool that the Roundtable developed is a personal computer-based information management system, used in developing an effective toxic chemical release reporting capability as required under Title III of the Superfund Amendment Reauthorization Act (SARA). Planning and developing such a system would cost the Roundtable in all about $55,000, according to ERM. However, by jointly developing the tool, the cost to each company is cut sharply and the resulting tool is much better because of the wide spectrum of input by Roundtable members' environmental specialists.

Changes in strategies on the environment must go beyond operations to deal with concerns even at the design and engineering stage.

In a report on the state of chemical engineering in the United States, the National Research Council noted that the "traditional approach to environmental concerns in process design—establishing the process and then providing the necessary safety and environmental controls—must give way to a new approach that considers at the earliest stages of design such factors as process resilience to changes in inputs, minimization of toxic intermediates and products, and safe response to upset conditions, start-ups, and shutdowns."

Bottom Line

• •

Petroleum companies will restructure in the 1990s over environmental concerns if for no other reason than the environment will have a very direct and extreme effect on their near- and long-term profitability.

An early sign of that was a total of more than $3 billion in writedowns in 1989 taken by Exxon, Chevron, ARCO, Phillips, Sun, and Texaco specifically in response to environmental issues. Exxon's related mainly to the Exxon

Valdez spill, Chevron/Phillips/Texaco to the plunging value of the Point Arguello project due to interminable permitting delays, and ARCO and Sun to more general provisions for future environmental costs. Expect more such writedowns in 1990–91, especially as the shape of new federal and state regulation and legislation becomes clearer.

Any company that has not done so would be well-advised to establish financial reserves for continuing and expected expenses related to environmental clean-ups. At the beginning of 1990, Amoco, ARCO, Chevron, Diamond Shamrock, and Texaco together set aside a combined $849 million in after-tax reserves for that purpose.

The costs identified thus far already are staggering. Crown Central Petroleum Corp., in testimony before DOE on the National Energy Strategy, estimated that new and upcoming major legislation in the federal government affecting the refining industry will incur compliance costs totaling $18.8–$22.2 billion/ yr. That does not include the billions spent every year to comply with existing federal environmental regulations, much less the proliferation of new state and local laws.

Other cost estimates boggle the mind. Unocal Corp. estimates the petroleum industry will spend as much as $50 billion on environmental controls in 1989– 93. That is more than double the total net earnings of the top 400 oil and gas companies in 1989.

Even that number is likely to prove conservative, Unocal contends.

In addition, the company estimates that industry spending for California's AQMP alone will total $130 billion in the 1990s.

Independent Petroleum Association of America (IPAA) sees environmental compliance as the single biggest cost center in exploration and production activities for U.S. independents in the 1990s.

Independent producers and refiners will have the toughest time in redirecting their corporate structures to accommodate environmental concerns. The smaller companies do not have the financial resources to increase staff, and generally they are stretched so thin on existing staff that they cannot designate many employees to take the point on environmental monitoring and compliance.

Independents are going to have to pursue cooperation and consolidation of efforts on environmental initiatives. The Roundtable approach espoused by ERM might be especially useful for independents in pooling their resources and reducing their costs in hiring environmental service contractors and consultants. This approach would be critical for independents in such high-cost areas as spill prevention and cleanup, produced waste handling, and wastewater problems.

What IPAA is pursuing in Washington that could prove a real boon to independents is an environmental tax credit. A tax credit for investments in environmental remediation and compliance might be the *only* new incentive that will survive among independent producers' requests for tax credits and other incentives from Washington in the 1990s. Given the problems over the deficits and the accelerating competition for a dwindling federal trough, even the small oil and gas producers are not likely to win favor on more tax credits. But, as is the case with the thousands of small service station operators threatened with extinction over new underground storage tank rules and concomitant unobtainable insurance, the indies can make the valid case that compliance without some sort of recognition in the tax code could mean extinction for them.

IPAA President H. B. "Bud" Scoggins points out, however, that even with an environmental investment tax credit, an independent still must deal with the problem of inadequate cash flow just to make the environmental investment in the first place.

"One of the biggest problems is just knowing what he's going to have to do," Scoggins said. "He can't just chase after it with a butterfly net."

Therein lies an opportunity not only for networking and pooling resources, but also for financial institutions and other investors. Given the proper tax code treatment, why can't a large, cash-rich institutional investor infuse money into independents' environmental programs in exchange for stock or some other yield mechanism? There are some busy little independents these days who would otherwise be out of business were it not for some large insurance companies and the like setting up limited partnerships to fund drilling programs. Why can't a similar approach work for environmental programs?

American companies' staggering investment in environmental mitigation and controls may have one unexpected benefit. If California truly is the bellwether on environmental issues, then what companies are spending now in the United States to comply with a California-style regimen, competitors overseas still have to look forward to, if efforts to internationalize environmental standards are successful. That could give American oil companies an edge in the next century, according to Unocal Chairman Richard Stegemeier.

"To assure a level playing field, however, we must work toward international standards, and quickly, so that all populations and all environments are equally protected," Stegemeier said. "This will be one of the most important challenges of the next decade."

Restructuring Approaches

· ·

Restructuring in response to environmental concerns in the 1990s will feature a wide variety of approaches.

Amoco Production Co., for example, has consolidated its environmental/safety/health functions as much as possible, with one department responsible for them above the plant level (Fig. 5–1). Six managers of environmental/safety/health areas report to Walter Quanstrom, vice-president for environmental affairs and safety. Four of those managers represent Amoco's principal operating companies, and the other two are with corporate staff, one specializing in industrial hygiene, safety, and compliance, the other in environmental conservation, product safety, and toxicology. Quanstrom reports directly to a board member.

Quanstrom offers this rationale for consolidation:

"We are trying to develop a critical mass of professionally trained environmental experts. They all get better if they are in a group. We have professionals in toxicology, pharmacology, chemical engineering, and operations safety. We have an inherent bias toward a large mix of individuals. It helps in cross-discipline training as well."

Phillips recently created the post of vice-president of quality, environment, and safety and realigned the department. Also reporting to that position are medical services, corporate management services, reviews and assessments unit, participative action teams, and the company suggestion plan.

British Petroleum Co. PLC has undertaken significant restructuring changes reflecting environmental concerns at the top. The company formed a four-member environmental audit committee on its board of directors consisting exclusively of outside directors. The purpose of the committee is to assess the company's efforts at restructuring to deal with environmental issues.

At the same time, BP is continuing to review all of its health, safety, and environmental quality policies. It also increased staff in each of those areas.

BP began a waste minimization program companywide, focusing not just on toxic and hazardous waste, but even waste paper in offices.

BP also formed an environmental research department at its Cleveland, Ohio, research and development center. It created the department by consolidating more than 30 scientists in a single area to work on the special needs and problems of environmental research.

That effort has already yielded results. The Cleveland research center has developed BP's best demonstrated available technology—required under EPA regulations—for cleaning oily wastes.

Conoco has maintained a strong emphasis on safety and the environment since the 1960s. It continues that tradition with a bold, sweeping series of environmental initiatives, including the controversial adoption of double-hulled tankers, announced in early 1990 (Table 5–1). Most recently, the company took its existing environmental affairs/occupational health department and combined it with a reconfigured safety department. It was an about-face for Conoco, which had decentralized those operations in the late 1980s, setting up a separate corporate safety department.

Conoco consistently has been one of the top two companies in terms of its safety record for more than a decade. However, the decentralization effort did not produce a change in its safety record. So the company decided to reconsolidate safety, environment, and occupational health under a single manager, Barry Kumins. He is on Conoco's Coordinating Council, which consists of 30–40 senior managers. Kumins reports directly to the senior vice-president of administration, who sits on Conoco's Committee of Six, the company's top decision-makers. The consolidation has already resulted in internal benefits, according to Kumins—improving communications with other segments of parent Du Pont Co. It also gave a heightened emphasis to occupational health, which had been hidden in a niche in the environmental department. In addition, Conoco has also begun a company-wide waste minimization program.

Du Pont in 1989 launched its own set of environmental initiatives for its agricultural products business in the 1990s—a program that would serve as a good blueprint for action by downstream petroleum businesses. They call for:

- Assessing new products for safety and environmental compatibility.
- Maintaining a continuing review of products for their environmental acceptability and sometimes phasing out products as more environmentally acceptable products come along.
- Reducing waste in manufacturing processes by half by 1995.

- Pursuing improved regulations related to how crop protection products are evaluated, registered, used, and monitored.
- Beefing up programs in safety and training education to ensure products are safely applied.
- Accelerating research in biotechnology to pursue a more environmentally benign way to protect crops.

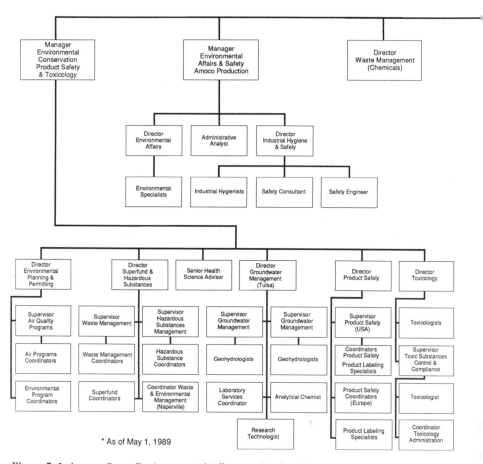

Figure 5-1 Amoco Corp. Environmental Affairs and Safety. *

ARCO in the late 1980s gradually realigned and consolidated all of its departments related to external affairs under government affairs. That included environment, safety, and health as well as government, public relations, and tax departments.

The rationale for the consolidation was that external concerns invariably return to government action—either through legislation or regulation.

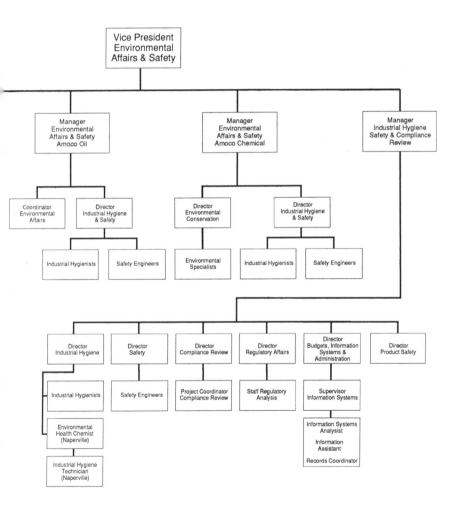

Table 5–1 Conoco's Environmental Initiatives

Cost: More than $50 million/yr.

Construction of only double-hulled tankers in the future (the first in the industry to issue this mandate).

More than a one-third reduction in toxic air emissions and hazardous waste levels from current levels by the end of 1993.

Watchdog citizen advisory councils to identify and implement plans and activities at U.S. and European refineries, major production facilities, and other major installations worldwide.

Double containment systems at all new and renovated company-owned gasoline stations.

Stepped up research into fuels that improve environmental quality, including reformulated gasolines and expanded use of CNG and propane in fleet vehicles.

Increased emphasis on specific environmental action plans in the compensation of all top managers.

Creation of an annual environmental fellowship program for outsiders to work with Conoco on environmental matters.

Accelerated emphasis on understanding environmental sensitivities involved in oil and gas development worldwide, notably a program focusing on developing hydrocarbons in rain forests.

Strive for 100% recycling of waste generated at Houston headquarters, to be expanded to London and Ponca City, Oklahoma.

ARCO's basic policy on safety and the environment has not changed in 20 years, according to Bill Leake, vice-president of safety and environment. Thus, it did not see a need to change the structure of the safety/environment department.

What has changed at ARCO, however, is a fundamental restructuring of the company's attitude toward environmental issues, according to ARCO Senior Vice-President Bill Dickerson.

That change in attitude was a response to the juggernaut of change looming on the horizon, notably with regard to air quality concerns, especially in California. With the catalyst of the Exxon Valdez spill, ARCO then came to the conclusion that the environment is going to be the overwhelmingly dominant issue of the 1990s.

Specifically, ARCO undertook new approaches in its thinking on the key issues of air quality and oil spills.

ARCO questioned the basic capability of the petroleum industry to respond adequately to catastrophic oil spills. It advocated instead the establishment of a national oil spill response and cleanup capability covering the entire United States and placed under the sole auspices of the Coast Guard. Under ARCO's plan, the petroleum industry would create a fund of at least $1 billion to furnish the Coast Guard with the capability of cleaning up all spills. Initially, that would involve start-up costs of $100 million and $35–$40 million/yr in operating costs.

That sort of approach, according to Dickerson, would clarify any misunderstanding over responsibility for spill response and cleanup. Confusion over responsibility was one of the chief aggravating factors in the Exxon Valdez spill. Further, ARCO contends its approach would avoid the perception of conflict of interest for companies involved in a cleanup. Exxon's assumption of responsibility for the cleanup in Prince William Sound has continued to haunt it as Alaska and federal politicians and the environmental lobby took particular glee in holding Exxon's feet to the fire amid disputing definitions of "clean."

ARCO does not take issue with the concept of Petroleum Industry Response Organization (PIRO), the petroleum industry's voluntary oil spill response and cleanup organization, set up shortly after the Valdez spill. The distinction here is that the public essentially will not trust the industry to be rigorous and thorough in cleaning up after itself. Since the Coast Guard would have inspection and approval authority, the reasoning goes, it might as well have the initial responsibility, with industry footing the bill.

ARCO has already taken ambitious steps toward coping with air quality concerns. It recognizes that the rest of the United States would follow California's lead in reaching a solution to the air quality problem.

"The direction things seemed to be heading is a push toward alternate fuels such as methanol or electric cars without any real research behind it," Dickerson said.

"It also became apparent that it did not make a difference to the public whether or not there *was* research behind it. The public wanted a solution, even if it was the wrong solution."

That recognition led to ARCO's push to reformulate gasoline. Developing the new, low-emission fuel, EC-1, is a perfect example of how the industry should respond to born-again environmentalism. ARCO recognized the problem as an opportunity and in September 1989 launched the successful new product. Faced with a costly, perhaps unmanageable "solution," ARCO not only solved the problem, but also carved out a new market niche for itself in California's retail gasoline market.

ARCO's first major salvo in the greening of the petroleum industry hit a bullseye. Not only is the company getting the significant emissions reductions it had sought in older cars, but it also is getting significant reductions in new model year cars, which was unexpected. ARCO has estimated that if every older vehicle in the South Coast air basin switched to EC-1, the region would achieve federal air quality standards called for under AQMP five years sooner. This kind of effort could dodge the alternate fuels bullet altogether.

The effort was not lost on Scaqmd, the AQMP's Dr. Frankenstein. The agency bestowed a special recognition resolution on ARCO for introducing EC-1. This is the same agency whose air quality plan calls for eliminating the petroleum industry from southern California early in the next century.

Fortune magazine named EC-1 one of its top products of the year, and the state of California called it the product of the year.

One of the best images in the industry for being environmentally conscientious belongs to Chevron. Its director of health, environment, and loss prevention, Lyn Arscott, was the first in that position to be named president of the Society of Petroleum Engineers (1988–89).

Chevron in 1989 reorganized its environmental and related functions at the corporate level. That entailed creating four major groups: a regulatory oversight group that works through trade associations and governments; a loss prevention group specializing in industrial hygiene, safety, fire, emergency response, and spills; a large toxicology lab; and a Superfund/waste management group.

Risk Assessment

· ·

Another change in Chevron's environmental policymaking deals with risk assessment.

Chevron set up a process hazards management group that seeks to manage risk through hazard/operability studies—finely detailed analyses of procedures and processes that call for speculative projections over a range of risks. Chevron prioritizes those risks, then reconfigures the process to reduce the level of risk.

Since risk assessment is still a nascent science, determining risk priorities can be a tricky business. But it's worth the effort, according to Arscott.

"Reducing process risk often increases productivity, and that in turn, reduces costs," said Arscott.

Chevron's philosophy on risk is "managing for assurance," which entails moving beyond mere compliance.

"Compliance is a given," said Arscott. "There are risks, and you may still be within legal bounds, but you still have to head off accidents. You've got to learn to manage risk with training and procedures."

Managing risk with training and procedures goes beyond company employees. Companies also should put more effort into ensuring adequate training of contractor employees, especially among the smaller contractors that might not have the resources to maintain the same level of risk awareness among its employees.

The key to managing risk in the 1990s is to instill in each employee a commitment to be responsible for his own safety and for the safety of his coworkers, contends Unocal Chairman Richard Stegemeier.

In a letter to employees in June 1989, Stegemeier wrote, "I'd like to remind every employee to be on the alert for potentially unsafe conditions and to bring them to the attention of your immediate supervisors. If adequate steps are not taken to correct an unsafe condition, you should alert management at a higher level—without fear of disciplinary action. I'd also like to remind you that loyalty to your fellow workers should never include ignoring violations of company safety policy that might put others at risk or result in serious environmental incidents."

Such efforts to inculcate the whistleblower ethic are likely to spread across the U.S. petroleum industry, especially in light of multibillion-dollar liability exposures such as the Exxon Valdez and the Phillips and ARCO Texas petrochemical plant blasts.

Unocal's accident reporting system reflects the high priority its senior management gives to safety. In the event of a lost-workday incident anywhere in the company, the written report not only is reported through normal channels, it also is sent directly to Stegemeier.

"In particular, I want to know why the accident happened and what steps are being taken to see that the same kind of accident doesn't happen again," he said. "I realize that reporting to the chief executive officers can seem intimidating, but it's a good way to send a clear signal to our people in the field that safety is top priority. Our goal is not to punish or to fix blame, but to determine exactly why accidents occur—so we can better prevent them."

Another aspect of Unocal's safety program emphasis is its safety audit program.

The company in fall 1988 began implementing its safety audit program company-wide. It provides for systematic inspection of facilities and a review of the strengths and weaknesses of their individual safety program. That and other safety programs enabled Unocal to cut lost-workday cases by more than 20% in 1988 from the year before.

The payoff from improved safety performance more than compensates for the investment. A study by Du Pont Management Services shows that the financial incentives for improved safety performances are stronger than ever. Data from 676 companies showed an average potential savings of $3.1 million per company during a five-year period from improved worker safety. For all the companies during that period, the total savings would be $2.1 billion.

The companies posted 35,100 lost-workday cases in the survey period at a cost of about $23,500/case, or about $825 million/yr in all, based on National Safety Council 1987 estimates.

Du Pont estimates that a sound safety management program could cut a company's potential lost workdays by 36% in the first year and by as much as 77% in the fifth year of the program. Those projections account for the individual company savings estimates. The Du Pont study covered a broad spectrum of companies, including chemicals, refining, food processing, electronics, and trucking.

Du Pont considers the elements essential for an excellent safety management program in any business to be:

- Safety policy and management commitment
- A structured organization for safety
- Safety as a line organization responsibility
- Safety audits
- Safety goals and objectives
- Standards of safe performance
- Supportive safety personnel
- Injury and incident investigations and reports
- Safety motivation and communications
- Safety training

Another way to manage risk is to identify it at an early stage, such as with a risk assessment.

Risk assessments are part of the remedial investigation feasibility studies that the EPA and potentially responsible companies must do in preparing a proper remediation plan for a Superfund site cleanup.

Those conducting a Superfund risk assessment must identify contaminants at a site and then select chemicals that are representative indicators, or those that pose 90% or more of the risks. Then, the assessor must conduct an exposure assessment to identify ways the chemicals are likely to affect a contaminated area and possibly move off-site to affect human health and the environment. In addition, the assessor must evaluate the toxicological properties of the indicator chemicals. That entails determining toxicity according to acute, chronic, carcinogenic, teratogenic, and environmental effects. Having done all that, the assessor then must estimate quantitatively the potential hazard from each indicator chemical singly and in combination.

Environmental Resources Management has developed a computer-aided risk assessment (CARA) program that the company contends can cut the time needed for such quantitative analysis by 90% and at the same time sharply reduce the likelihood of error by 8–9%. Risk assessments that used to take days or even weeks can be completed in just hours with CARA, ERM claims.

Another part of the process in reducing a company's liability involves environmental audits. These tools are becoming increasingly important for petroleum companies in light of the massive consolidation that began in the latter half of the 1980s. Companies seeking to streamline operations have set about disposing of marginal properties or other assets. With the concern over accountability that arose after EPA's Superfund program identified so many "orphan" waste sites—in which not all of the parties possibly responsible for polluting a site can be identified—liability concerns have prompted financial institutions to tighten the rules. It thus becomes very difficult to obtain financing for any property transactions without first conducting an environmental audit.

The People Factor

• •

Petroleum companies, like those in other industries, are fond of saying that their employees are their most important resources.

The new high profile for environmental accountability means that no company can merely pay lip service to that slogan any more. Just ask Exxon.

Chevron's Arscott stresses imbuing all operations people with a keen awareness of compliance with regulations and protection of the environment.

"You have to get the line manager involved. You cannot run a program from headquarters. You have to get risk awareness all the way down the line to the wellhead," Arscott said.

That also involves developing a strong sense of public affairs awareness among operations people on environmental issues, Arscott contends.

"The line operator or plant manager can't isolate himself from the community's concerns. A lot of engineers have had to be retrained in this regard."

What is especially crucial is that the person directly responsible for environmental policy must have the full weight of corporate power behind him. Given the greater likelihood of environmental violations netting criminal penalties, reason dictates that culpability will accrue beyond operations people to managers.

Perhaps the knottiest personnel problem facing petroleum companies is the critical shortage of qualified people to fill positions related to environment, safety, and health.

Exacerbating that situation, according to Hart Environmental Management Pres. Fred Hart, is the continuing bleed-off of talent into government and government-contractor work.

"We need to lessen reliance on the government for studies. There is already too much redundancy, inefficiency, and waste."

Hart claims his firm can perform remedial investigations for about one third the cost of one conducted by the government or a government contractor.

"That's because we have better qualified people, and we are not required to get as much data, so it doesn't take as much time," Hart said.

There are differing approaches among petroleum companies in dealing with the shortage of qualified environmental personnel.

Smaller companies in particular are likely to take a skilled person who might otherwise be laid off because market conditions don't merit a certain staff level at that position, and retrain him in an environmental specialty. Not only would that person be broadening his scope at the company, but his primary skill would be "banked" for when future demand warranted it.

ARCO is not interested in that approach. The company focuses on recruiting skilled, licensed, certified professionals in the environmental/safety/health arena, despite the fact that competition for these skilled people is very keen these days.

Some companies may have regretted their recruiting efforts in the early 1980s, when a boom oil market and resulting shortage of qualified petroleum scientists led some to recruit engineers in other fields and retrain them in petroleum engineering.

Screening for the right kind of professionals and technicians to step into the role of environmental specialists will remain a challenge for petroleum companies.

It may be wise to emulate the rigorous approach practiced by Law Cos., an Atlanta geotechnical engineering firm that specializes in assessing environmental hazards and devising solutions for them. Law requires that all of its engineers take specialized courses. After that, each engineer is called before a panel for a four- to six-hour interview, a sort of master's orals. If the engineer—his number or weight of degrees notwithstanding—fails the exam, Law won't register him for work as an engineer.

Law Chairman R. K. Sehgal notes, "Sometimes, we find someone with a great deal of book knowledge who is, unfortunately, unable to tell us how equations and theories and the like can be applied in a pragmatic, client-oriented manner. Such a person will fail our examination . . . and probably, eventually leave the company."

Nevertheless, industry is likely to look more to academia for its future supply of environmental specialists. Petroleum companies can do some informal campus recruiting while they fund valuable research on environmental problems.

At the same time, managers should begin efforts to develop their own environmental expertise, says John Baden, head of the Foundation for Research on Economics and the Environment, Bozeman, Montana, and Seattle, Washington.

Baden, noting that no executive business or MBA program effectively prepares management for making discriminating choices on environmental matters, offers his own suggestions for a business and environment program as an option for MBA programs.[1]

Baden recommends executives take courses in history of environmental/natural resource policy, environmental/natural resource economics, environmental ethics and philosophy, and the political economy of environmental policy.

Chevron's Arscott sees the environment/safety/health area as the pathway for future corporate stardom.

"You have two options. There is the default option, in which you take a good, productive person, who perhaps is not doing too well in his current job and retrain him," he said.

"Or, you take one of the best people, someone with high potential. If you see him as a possible future leader in the industry, that employee should have had some experience in the environment/safety/health field."

The generation of top managers in the petroleum industry today has become heavily infiltrated with executives coming from financial backgrounds, due to a need for their skills in coping with restructuring and devastated oil economics in the 1980s.

Correspondingly, many of the 1990s leaders in the petroleum industry will bring with them expertise in environmental and related issues.

Carving New Niches

• •

The most obvious effort to carve out new niches or create new profit opportunities in response to revived environmental concerns is the bandwagon that ARCO started.

Other refiner-marketers have quickly followed suit, which will be detailed in an upcoming chapter.

The new, "environmentally preferred" products won't be limited to gasoline. Shell, which also is marketing a low-emission reformulated gasoline in selected U.S. markets, introduced a reformulated heavy-duty motor oil. Shell's reformulation involves an extended-life additive that is designed to improve engine cleanliness and protection and better control oil consumption. Tests show the new motor oil outperforms its previous formulation while reducing the effects on the environment.

Such efforts are in keeping with a joint cooperative program between the oil industry and Detroit's Big Three automakers to develop reformulated gasolines and the vehicles compatible with them. That effort won't come cheap. ARCO, for example, plans revamps at its Cherry Point, Washington, and Carson, California, refineries at a cost of $1 billion each to produce the reformulated fuels it thinks will meet federal air quality standards. Whether some of California's more extreme air emissions standards are achievable by any petroleum-based fuel will remain a question mark for some time to come.

Petrochemical producers also will be carving out new product niches to accommodate concerns over less environmentally acceptable products. Chemical producers already are developing replacements for CFCs in refrigerants and aerosols, nitrogen-based fertilizers, nonbiodegradable plastics, some toxic agrochemicals, and paints and coatings with high emissions of VOCs. In the United States, the amount of CFCs used in insulating foam—a petrochemical accounting for 80% of the CFCs used in a refrigerator—has fallen by half recently.

A thornier problem is developing viable alternatives to the CFC cooling fluid freon, none of which have proven to be as efficient. One oil company subsidiary has put together a recycling and consumer education program to deal with the freon problem. IG-LO Products, a unit of Valvoline Oil Co. and the leading U.S. marketer of automotive refrigerant, began establishing freon recycling units at service stations to capture fugitive freon emissions and allow them to be reused. IG-LO also is eliminating inexpensive do-it-yourself freon injection kits, thus avoiding widespread misuse. Instead, it is introducing deluxe freon diagnostic kits at fivefold the cost aimed at the more upscale—and presumably more environmentally conscious—consumer, thus sending most users to service centers that have the recycling units. In addition, the company launched a $1 million consumer education program designed to promote proper use and recycling of freon.

Petrochemical producers are also getting an unexpected windfall from development of new markets for products made from recycled plastics—lumber, furnishings, highway markers and signs, construction materials—that are backing out more expensive and less durable materials.

Diversifying into environmental services is a natural outgrowth for petroleum companies, especially the larger ones.

Most sizable oil and gas companies have had to increase staff in this area in recent years while maintaining or cutting other staff levels.

More companies will recognize value in marketing that in-house skill throughout each segment of the company, and some, even outside the company.

Companies such as Amoco, BP, and Phillips have ventured into waste management services through acquisition or start-up.

These efforts largely were born of the industry's own intensified efforts at waste management in the wake of the proliferation of zero-risk standards and soaring costs for end-of-pipe solutions for toxic and hazardous substances and increasing costs of landfilling and waste incineration for conventional substances.

Phillips contends that waste minimization is its single best environmental practice, not only because it helps to solve environmental problems but also because it is the cheapest way to efficiently handle materials and reduce capital expenditures.

Amoco, which acquired Waste-Tech Services Inc. from Bechtel National Inc., sees the acquisition as a logical extension of its capabilities in hazardous waste processing. Before the acquisition, Amoco operated two hazardous waste incinerators and had extensive experience in the field. Waste-Tech provides on-site and regional incineration services to industry for destruction of hazardous wastes. It is involved in the design, permitting, construction, operation, and ownership of incineration facilities.

Oil companies getting into the waste management field are going to have their hands full competing with some giant firms that have been in the business for decades.

Industry is pursuing waste management even as a new source of energy supplies. Gas Research Institute and the Institute of Gas Technology are participating in a Department of Energy experimental demonstration project to convert municipal solid waste to methane that is under way at Walt Disney World in Orlando, Florida. Leading the venture is DOE's Solar Energy Research Institute. It involves converting Disney's experimental wastewater treatment unit, which normally processes sewage sludge, to handle solid wastes. The experimental test unit at Disney involves anaerobic digestion—essentially the process that occurs naturally in landfills—to convert solid wastes to low-cost methane and carbon dioxide. DOE projects in the past have successfully duplicated this process in dedicated chemical reactors. The SERI/GRI project focuses on improving yield and economics.

Amoco does not want to venture into environmental services beyond the Waste-Tech acquisition. Like Chevron and ARCO, Amoco sees environmental services as essentially an in-house consulting service dedicated to one client.

Conoco, on the other hand, sees some real growth potential in considering in-house environmental services as business opportunities. One reason, the company figures, is the increased activism among states at the environmental regulatory level.

Conoco has launched its own maiden effort in that regard, forming an environmental geosciences group to provide support for parent Du Pont's environmental projects.

The new group consists of 11 geoscience professionals from Conoco's worldwide exploration department. It provides geological, geophysical, and mapping services to other operating areas within Conoco and other Du Pont units.

The idea is to provide a wide array of geoscience services to support accelerated environmental mitigation programs, notably for projects affecting soil or groundwater.

The environmental geosciences group will handle on-site guidance during drilling of sample boreholes and monitoring wells, groundwater and soil sampling, sample analysis, geophysical measuring, gravity and magnetic surveying, site characterization, computerized 3-D subsurface mapping, and computer modeling. Along with assistance to company operators, the group will also help provide statistical data that are mushrooming in light of increased state and federal regulations.

Du Pont itself has leaped full bore into environmental services as profit center. It created a new unit, Safety and Environmental Resources, to manage industrial safety and environmental programs. The focus is on waste management, safety and training, and environmental remediation. Du Pont expects the unit's annual revenues to climb from about $100 million in 1990 to $1 billion at the end of the century.

Several areas lend themselves to petroleum companies developing and marketing new products and technologies in response to environmental concerns.

One company formerly in the oil and gas business, Yankee Cos., has restructured to concentrate on the environmental cleanup business. It has produced a neutralizing agent that prevents spilled oil from sticking to gravel or sandstone. Another Yankee product—obtained via acquisition of the company that invented it—solidifies oil by encapsulating it. Yankee tested both at the Exxon Valdez spill and reported good results.

Bioremediation is certain to be an attractive R&D area for yielding marketable new products and services in the environmental arena, especially given the high profile use of this environmentally benign approach in the Valdez and Mega Borg spill.

The Louisiana State University Institute for Environmental Studies is pressing research into microbial bioremediation to clean up hazardous waste sites. EPA in 1988 approved plans to make the Old Inger abandoned hazardous waste site near Darrow, Louisiana, the first Superfund site in the United States to be cleaned up using microbial bioremediation. LSU's Center for Energy Studies, which helped fund the research, estimates the cleanup would cost about $8 million with microbial bioremediation versus as much as $25 million for burial and incineration—each of which carries its own environmental impact burden. Microbial bioremediation tests at LSU also have involved polyvinyl chloride,

PCBs and dioxin-tainted PCPs, petrochemicals, creosote, PAHs, oily waste, and several pesticides.

There are new approaches on remediation that could save petroleum companies money. ENSR Corp. developed a "fast-track" remediation scheme for the Swope Oil & Chemical Co. Superfund site in New Jersey. ENSR's plan, hailed by EPA as a model approach, involved breaking the overall cleanup design plan into tasks by order. Then ENSR submitted contract documents for 17 individual activities, each with its own design, quality assurance, and health and safety plan. With permitting agencies reviewing and approving task by task, ENSR did not have to wait for the entire plan to be approved before starting cleanup.

Search for Solutions

• •

What kinds of solutions should the petroleum industry be searching for in dealing with environmental problems?

Some may be far removed from the industry's usual frame of reference—calling for a sort of technological or intellectual diversification.

For example, consider the herculean effort required under the AQMP for cleaning up the dirty air over Los Angeles. The problem is not just the source of emissions, but the metereological conditions and the topography of the Los Angeles basin. Because of the nature of the basin's "closed bowl," pollution gets trapped in the lower atmosphere by a temperature inversion. That is compounded by the lack of precipitation in the desert region, leaving a situation whereby pollutants achieve a concentration many times their rate of emissions. The air pollution does not get cleaned up until rain washes it away. But you can't readily transform a region's terrain or weather patterns, right?

Scientists with Hughes Aircraft Co. have proposed a mountain-to-Mahomet type of solution for Los Angeles's vexing air quality problems. Because cool land air over Los Angeles essentially blows out to sea to replace warmer sea air during the night (reversing the process during the day), why not create a giant curtain of water to scrub the atmosphere as it circulates out to sea, the Hughes scientists proposed.

They would mount 40 water jets on the seabed to spray a fountain of water 2,000 ft in the air at night. Pressure to generate the jet stream of water

126

would be accumulated during the day through wave action. On rainy days, the wave power could be diverted to generate electricity. The weakened waves would in turn cut beach erosion by as much as 60%. The Hughes scientists estimate this coastal air-wash system would cut air pollution in the Los Angeles basin by 53% in the summer and 79% in the winter. Cost would be one-tenth that of the AQMP plan for achieving the same amount of pollution reduction, they claim. Scaqmd rejected the plan on environmental impact grounds.

I am not advocating this as a viable proposal to resolve industry's dilemma over the AQMP. But this kind of approach should serve to illustrate that the petroleum industry needs to move beyond a conventional frame of reference, even outside the industry, in considering solutions to environmental problems.

Maybe the real solution to auto emissions is not reformulated or alternate fuels, but the perfect tailpipe filter that captures the very nanogram of pollutant without affecting vehicle performance. So perhaps petroleum companies looking at the prospect of being driven from the third biggest gasoline market in the world should consider investing in research in auto exhaust systems design, or maybe giant offshore water fountains. Industry must adapt to the new environmental movement by keeping itself open to creative, innovative solutions.

One thing that is needed is for industry and government to break the circle of the short-term environmental fix, maintains Monsanto Co. Chairman Richard Mahoney. The U.S. approach to environmental protection is flawed by its characteristic impatience, he contends.

"We want our environmental problems fixed and fixed now. Instant gratification is a natural desire, but it causes Congress and state legislatures to focus upon the short-term fix, one short step at a time."

Consequently, laws and regulations dealing with environmental problems are adopted piecemeal, typically resulting in an end-of-pipe solution.

"In the legislative stampede to the short-term fix, it didn't matter whether or not these were the best approaches, whether they were whole solutions, or whether in fact the solutions could cause more problems," Mahoney said.

That in turn creates more public dissatisfaction with what it perceives as a lack of progress on environmental problems.

"Broadly speaking, what we must have are approaches which encourage voluntary and creative initiatives and market incentives to do the right thing—and to do it faster with real results," Mahoney said.

He called on Congress and regulatory agencies to shift their focus from legislating rigidly prescriptive rules to developing policies that would encourage voluntary, market-based initiatives across American industry.

If anything, such voluntary initiatives are stifled by Washington, Mahoney contends. He cited Monsanto's own voluntary commitment to a goal of zero toxic air emissions with a benchmark reduction of 90% by yearend 1992. Yet under pending regulation, several major Monsanto facilities would not qualify for regulatory credit for reducing air emissions by 90%.

Mahoney feels that it is up to companies to commit themselves to the idea of having no significant impact on the environment, opting to eliminate, reuse, or recycle waste on-site.

This upsurge in environmental concern will give rise to a new breed of "ecopreneurs" in the 1990s, akin to the entrepreneurial boom of the 1980s, that will attract much of the available venture capital.

It is important to remember that new environmental standards will create business opportunities, perhaps whole new operating segments and profit centers. As petroleum companies respond to the newly resurgent environmental consciousness, they should think in terms of investment opportunities and new business start-ups related to the activities they must undertake to respond to environmental concerns.

The petroleum industry can grab a piece of that burgeoning new market—not to mention ensure its economic survival—if it is willing to demonstrate leadership on environmental issues as opposed to its traditional reactive stance.

Gary Dillard, Shell vice-president for public affairs, noted the blizzard of federal and state legislation being considered at the moment and the fact that some of that legislation will become law.[2]

"We in industry can rail against it all we like, but we must realize that we brought much of it on ourselves. If we had communicated more openly with the public all along, if we had shown more environmental leadership, we wouldn't be forced to fight some of this legislation now.

"In the worst possible scenario, we could lose control of our business. We could end up just like the nuclear power industry, which has been paralyzed by public fear and is now being managed by the public process."

Dillard contends that the first step in avoiding that kind of disaster and regaining the public trust is to improve environmental performance by going beyond mere compliance.

If we wait until the regulations are passed, if we wait until deadlines are upon us, then we comply by meeting minimum legal requirements, we forfeit any claim to leadership.

To be perceived as an environmental leader, we must set our standards above the minimum required and meet them all the time.

We must make a commitment to go beyond the letter of the law, to do what we know is right.

It's like having a savings account for difficult and unforeseen times; we have to build credibility and trust to have a positive image in the bank. We must instill that commitment in all our employees, and we must communicate it to the outside world.

Take the Initiative.

References

1. John Baden, "Save the Environment Without Destroying Your Profits," Manager's Journal, *Wall Street Journal,* August 21, 1989.
2. Gary Dillard, "Knowing Isn't Enough," *Shell News,* February 1989.

6

NATURAL GAS: FUEL OF THE FUTURE

Whatever the enormous challenges and pitfalls awaiting the petroleum industry in the Decade of the Environment, there is little doubt about one thing: natural gas has a bright future in the 1990s.

Growing demand for natural gas—in large part due to a rapidly accelerating shift from other energy sources on environmental grounds—together with the removal of regulatory constraints that have long hamstrung this abundant American resource may be the main source of the U.S. petroleum industry's optimism in the 1990s.

With market forces and consumer preferences determining the role of natural gas in the U.S. energy mix for the first time in 50 years, U.S. petroleum companies increasingly are shifting their focus to gas in favor of oil.

That emphasis on gas will continue, even as natural gas prices rise—because those price hikes are not likely to be sudden or steep over a sustained period of time, and because environmental and energy security concerns together dictate that natural gas is an energy source whose time has come.

There will be some environmental concerns for the natural gas industry, but they pale in significance in comparison with those for the rest of the petroleum industry. For example, the gas industry accounts for only 5–8% of world methane emissions, in turn making it a contributor to no more than 1–1.5% of the postulated greenhouse effect. These figures disregard the reductive influence the gas industry has on the greenhouse effect through utilization of coalbed or landfill methane.

There are some modest inroads natural gas will make in traditional oil markets, such as transportation fuels. Natural gas vehicles can cut emissions of carbon monoxide by as much as 99%, nitrogen oxides 65%, and reactive hydrocarbons 85%. Fleets that have shifted to natural gas have realized a significant savings, not only in fuel costs, but also in vehicle maintenance and repair. However, because it is likely to be limited to fleet vehicles, compressed natural gas as a motor fuel is not likely to amount to more than 5% of the overall gas market in the 1990s.

And natural gas demand will continue to grow in industrial and space heating and cooling sectors—albeit modestly as efficiencies improve.

But the spectacular growth in natural gas demand in the United States will occur in the electric power industry.

And the gas industry has the environmental movement to thank for that.

Brownouts and BTUs

· ·

Demand for electric power in the United States is expected to outstrip the overall rate of energy demand in the country in the 1990s.

The U.S. Department of Energy's Energy Information Administration (EIA) in 1987 projected that total U.S. energy consumption would grow at an annual rate of 1.3% between 1987 and 2000. By contrast, EIA forecasts, U.S. demand for electricity would jump 2.4% annually during the same period.

Growth in electricity demand will account for more than half the increase in total end-use demand for energy in the United States in the 1990s. Electricity generation is expected to grow from 36% of primary energy consumption in 1987 to 42% in 2000, according to EIA.

Electric utilities are expected to meet the growth in power demand by completing more than 44 gigawatts (GW) of capacity under construction or announced. Most of that construction will occur in 1991–95, projected EIA. Beyond that, utilities must construct an additional 32–73 GW (base case of 53 GW) of capacity that is currently unplanned. Most unplanned construction, accordingly, will occur in the latter half of the decade.

EIA figured in 1987 that most of the unplanned generating capacity will come from either pulverized coal steam plants (15 GW) or gas-fired/combined-cycle plants (22 GW) built as peak-load turbines. Cogeneration power and independent power producers are projected to add almost 30 GW of capacity in the 1987–2000 time frame.

EIA in 1987 figured electric power's fuel mix will break out by 2000 as coal 53%, nuclear 17%, hydro 9%, and oil/natural gas the remainder. The

132

agency estimates oil and gas fired plants' market share will rise from 15% in 1987 to 21% by 2000. By 2000, the combined percentage of generation from oil and gas is expected to exceed that from nuclear power. EIA pegged the total consumption of oil and gas by electric utilities at 3.4 million B/D of fuel oil equivalent from about 1.8 million B/D in 1987.

Bear in mind, these estimates were made shortly after the oil price collapse and before The Year the Earth Spoke Back and the Exxon Valdez.

Even in 1988, concerns were arising among the utilities that a capacity shortfall might be in the offing.

In October 1988, the North American Electric Reliability Council (NERC) warned that "several forces creating risks to the future electricity supply in the United States and Canada must be carefully managed."

The alternative? Soaring costs, draconian conservation measures, controlled brownouts.

"These forces that affect the reliability of electrical supply include a higher than expected growth in the demand for electricity, an imbalance between economy and reliability considerations, public policy disincentives, transmission access and deregulation issues, and environmental considerations," NERC said.

"Several geographic areas where electric system reliability is at risk are identified . . . Emergency measures, including controlled interruptions of customers' supply, are possible in these areas."

Underpinning those projections of electric power demand growth are expectations of a healthy economy continuing.

In a 1989 study on North American electric power trends, Arthur Andersen & Co. and Cambridge Energy Associates cited the linkage between economic growth and electric power demand growth.[1]

The study noted that from the end of World War II until the early 1970s, every added U.S. GNP dollar, in constant dollars, inevitably came in tandem with more than 1 kw-h of electric power sales. That changed dramatically in the 1970s, when the ratio of added kw-h to increased GNP dollars fell by about half.

But the oil price collapse changed all that, restoring the benefits of low-cost energy to the U.S. economy. In 1987 and 1988, electricity sales outstripped GNP growth by about 1.5%/yr, according to Andersen/Cambridge.

"If this is evidence of another inflection point in the relationship between electricity sales and economic activity, it has extremely important implications for the utility industry," the study concluded.

"If the pattern that persisted from World War II until the mid-1970s has reappeared, then significant new power plant construction will be needed in the 1990s to avert regional power shortages."

More importantly for power planners, the linkage between economy and power demand remains unbroken, even after energy crises.

Although lagging electric power demand from the mid-1970s to the mid-1980s has been blamed on soaring electricity costs, there is reason to believe that a more significant factor in the falling rate of growth in electricity demand is the stagnant economy that persisted during the same period. That begs a chicken-or-egg proposition, of course, since the same spiraling oil and gas costs that contributed so much to the spike in electricity costs also contributed to depressing other areas of the economy.

If the economy remains healthy and summers prove to be hot and dry in the 1990s, the need to build new power-generating capacity in the United States will take on a much greater urgency.

Although there was much surplus generating capacity in the 1970s and early 1980s, that won't be the case in the 1990s. Already, annual capacity margins are at their lowest point since the early 1970s. In some parts of the United States, annual capacity margins are at their lowest levels since World War II.

A study by the American Gas Association in 1989, using NEC data, found that at least 50,000 MW will be needed by 1997 in addition to the capacity currently under construction if peak electricity demand grows at 1.9%/yr. If peak demand grows at a rate of 2.9%/yr, which is a conservative estimate considering current trends, the U.S. electric power industry would have to construct almost 105,000 MW of capacity in addition to plants already under construction.

In 1988, the electric utility industry had 73,400 MW of capacity additions on the drawing board for the following 10-year planning period. That compares with 250,000 MW of planned capacity additions by utilities for the 10-year period beginning in 1979.

For the 1988–98 planning period, planned coal capacity additions are down 86% compared with the previous planning period. Planned nuclear capacity additions are down 85% in the same comparison.

In large part, that situation can be blamed on the environmental movement. Certainly, conservation has played a role in down-scaling planned capacity additions. And the overbuilt capacity of the late 1970s and early 1980s has understandably made utilities much more cautious in planning capacity additions.

But barring some draconian new conservation push nationwide or economic collapse, there simply will not be enough electric generation capacity to meet peak electricity demand in the mid-1990s unless more capacity additions make it onto the drawing boards.

Some see widespread power shortages and blackouts more commonplace as electricity demand continues to grow in the 1990s. The areas most likely to be affected are Florida and the New England, Mid-Atlantic, and North Central regions. A study by Salomon Brothers in 1989 pegged the shortfall at more than 100,000 megawatts, perhaps as much as 400,000 MW.[2] Some mostly completed nuclear power plants are being held up by intervenor actions, exacerbating the situation.

The shortfall could come even sooner than any had expected a few years ago, according to the Salomon Brothers study. The analysts looked at electric utilities' efforts to maintain a reserve margin—committed total capability capacity minus peak demand—of about 20%, which is derived from a probability analysis done by each utility to minimize the risk of a blackout to a 1% chance in 10 years.

In calculating utilities' aggregate capability, Salomon Brothers factored in planned nonutility power sources, including cogenerators, resource recovery plants, and independent power producers. The analysts used four scenarios of demand growth, assuming no demand growth in 1989, followed by, respectively, 2% growth each year thereafter, 3%, 4%, or 5%. Under the 2% growth case, reserve margins will fall below 20% in 1993; under 3%, 1992; under 4%, 1991; under 5%, early 1991. Even under the most conservative demand growth scenario, Salomon Brothers estimates utilities will have to add 107,700 MW to their capability in the 1990s. Correspondingly, that calls for 196,600 MW needed with 3% growth, 294,700 MW with 4% growth, and 402,600 MW with 5% growth, the study concluded. It also cautioned that the shortfall will be greater and sooner than under these projections if the nuclear plants under construction never start up.

Another development likely to further aggravate the situation is the fundamental change that is occurring in the structure of the electric utility industry. The Investor Responsibility Research Center (IRRC) in 1988 issued a report that concluded that the basic conditions were in place for the largest wave of mergers and financial restructuring activity the electric utility industry has seen since the 1930s.[3]

The culprit?

"Financial problems associated with nuclear power plants are proving to be the single most powerful catalyst for industry reorganization," the IRRC report said.

There are other contributing factors as well, but the real dilemma for electric utilities is how to add capacity at a controlled cost. Yet if anything, the lead times for construction and permitting of big baseload coal and nuclear power plants are getting even longer.

Building a large coal or nuclear power plant today brings into play concerns over environmental and safety regulations, long construction lead times, and public utility commissions disallowing certain outlays deemed imprudently incurred.

Three Mile Island set the Nimby syndrome in concrete for the nuclear power industry. And the CAA—especially a tougher version of it likely to be implemented in the coming years—could ensure a similar scenario for coal-fired power plants. Oil might have been an alternative, had it not been for the Exxon Valdez, soaring import levels, and to some extent, air quality concerns.

At least two-thirds of state utility regulatory bodies now require a least-cost approach to planning future load growth, in order to rein long-term electricity costs.

As the industry undergoes a wave of restructuring and consolidation, utilities will become even more conservative in controlling costs of capacity additions. Survivors of the shakeout will be held more accountable for financial performance, and will be less likely to be bailed out by public utility commissions on cost passthroughs of "imprudent" costs.

The Changing Regulatory Climate

Changes sweeping across the regulatory horizon of the electric power and gas industries generally will have a salutary effect on gas sales' incursions into the power market.

NATURAL GAS: FUEL OF THE FUTURE

Many of those changes will provide a definite boost to natural gas, and would have, even without the rebirth of environmentalism.

The jitters electric utility planners have developed in recent years are justified. Utilities and their investors can no longer presume they will be allowed to recover their full investment in power plants. An increasing trend among regulators has been to disallow returns on investments in what was later deemed to be excess capacity.

When planning big baseload plants, where the economics of nuclear and coal would hold sway, it used to be acceptable to sin on the side of largesse. No longer. Utilities now must use existing capacity, nonutility purchases, conservation, load management initiatives, and power imports to meet their needs.

That may work—until the mid-1990s. But the long lead times for nuclear, and increasingly, for coal, tend to preclude those fuels for near-term planning purposes.

So it follows that as utilities seek to match planning with incremental growth in demand, they will turn increasingly to smaller, modular power plants and purchases from cogenerators and independent power producers (IPPs). Most of those will be fired by gas. Gas-fired plants tend to be smaller, have much shorter lead times, and still burn a fuel whose price is relatively low.

Gas can be used in a variety of power plant schemes: steam plants, combustion turbines, combined cycle plants, and cogeneration plants. Moreover, gas-fired turbines are seldom larger than 100 MW, and combined cycle gas-fired plants are usually smaller than 200 MW. That kind of scale enables utilities to add new capacity when they need it, rather than commit massive outlays to a big baseload plant years in the future that could become a white elephant if demand falls short of projections. The modular approach eliminates the need for costly custom-designed engineering and labor-intensive construction programs.

Regulatory changes within the gas industry are paving the way for a greater market share for gas in power generation. In 1987, Congress amended the Powerplant and Industrial Fuel Use Act to relax restrictions on using natural gas for nonpeak electricity generation.

One of the biggest incremental new markets for gas in the United States has been the meteoric rise of cogeneration. Although Congress in 1978 passed the Public Utility Regulatory Policies Act (PURPA), it was not until the mid-1980s that its desired effect was seen. Court challenges and hindrances by public utility commissions stalled PURPA's intent, which was to give nonutility power generators a secure market by requiring electric utilities to buy power from them at the avoided cost rate. How one defined avoided cost made all the difference for the feasibility of cogeneration. The early, loose interpretation by some regulators essentially involved the avoided cost of having to build the same grassroots capacity from the ground up.

With the projected costs of power plants skyrocketing, that pushed avoided cost schemes up to premium levels and triggered a cogeneration boom that was first spawned in California, where the public utilities commission and the major utilities encouraged cogeneration growth. That boom spread, notably to the Texas Gulf Coast, where the vast refining and petrochemical complexes found the double benefits of producing huge volumes of process steam locally, as well as providing a secondary profit center in power sales.

However, the cogeneration projects proliferated to the point where the utilities were faced with being forced to buy more power than they could use, perhaps finding themselves eventually in the bind of having to shut down existing baseload capacity in order to buy the cogeneration power. Regulators rescued the utilities by tightening the parameters for avoided cost sales, focusing on the avoided cost of burning competing fuels. Consequently, the economics of cogeneration projects shifted from sales of overpriced power sales to their thermal value. That has resulted in a shakeout of cogeneration projects and developers.

IPPs are not without some of their own economic hurdles as well. Such projects generally depend upon project financing—debt secured solely by revenues from a single plant—which tends to be costly and complicated. Some projections estimate that 40% of proposed IPP capacity will be delayed and 20% will never be built, according to engineering consultants Burns & McConnell, Kansas City, Missouri.

And IPPs are not always the most economic choice from an investment standpoint, contends Burns & McConnell. The consultants cite their project for Old Dominion Electric for two new 400 megawatt coal fired units. They received 15 proposals from conventional contractors involving utility ownership and from IPPs. The conventional utility ownership alternative was 15–20% more economic on a life cycle basis than its nearest IPP competitor, according to Burns & McConnell.

The consultants also raise the question of reliability.

"It may not be politically popular, but public commissions should allow utilities to consider the reliability of IPPs in their long-term plans," Burns & McConnell said in a 1989 report.[4]

"Can we really depend on these plants being financed, permitted, and built on schedule, and to operate reliably for 40 years?"

Still, the deciding factor is likely to come down to cost, and IPPs can add coal-fired electric capacity for a fraction of conventional baseload utility plants.

And, the economics of cogeneration are very healthy today, partly because of its inherent energy efficiency and the low cost of its most attractive fuel: natural gas.

Because non-utility power generators were exempt from Fuel Use Act restrictions on burning natural gas prior to 1987, more than one-third of non-utility power generation is already fueled by gas. Low prices ensure that significantly more nonutility power generation will be fueled by gas in the future. Not only is the cost of gas as fuel low, the capital cost of gas-fired plants is low compared with the alternatives. The price of natural gas delivered to electric utilities has fallen in constant dollars since 1982. A slow decline in that price occurred in 1983–85, followed by a steep drop in 1986, before continuing a slow slide again.

The fall in gas prices has as much to do with regulatory reform, particularly by the Federal Energy Regulatory Commission, as it has with the collapse of oil prices in 1986. Decontrol of gas supplies, gas wellhead price decontrol, open access to transportation on pipelines, unbundling of pipeline services, market-indexed gas pricing contracts, improved ratemaking flexibility, and trimmed regulatory constraints, all helped to foster gas on gas competition, which has reined prices.

Deregulation initiatives in the gas and power industries have helped fuel spectacular growth—with more to come—in independent power generation, notably fired by gas in cogeneraton and conventional turbine plants.

Jean-Louis Poirer, in a 1989 study for RCG/Hagler, Bailly Inc., Washington, D.C., estimates that a total of 30,500 MW of IPP capacity will come on line by 1995. That will mushroom to as much as 90,000 MW by the end of the century, he projects.

Of the 42,760 MW of IPP capacity under construction or in development in early 1989, cogeneration accounted for 71% of the active capacity and almost half the active projects, Poirer found. Further, natural gas accounted for almost 29,000 MW of the active IPP capacity, of which 40% was on-line at the time.

Some IPP developers have projected IPP capacity nationwide reaching 200,000 kw by 2000, split 50–50 between load growth and replacement of existing baseload capacity.

Environmental constraints on nuclear-, coal-, and oil-fired electric power will ensure a built-in premium for natural gas continuing even as surging demand growth and declining deliverability boost natural gas prices again in the 1990s. Gas prices might double in the 1990s, as some predict, but the added costs of scrubbing equipment and uncertain economics for nuclear will likely offset higher gas prices. To the extent that nuclear power can make its case for safety and public acceptability and coal can make its case for economically viable clean technologies, that will determine the strength of natural gas in the power generation market in the United States.

The Case for Nuclear

As might well be expected, the nuclear power industry is seizing the opportunity to make its case—or rather, remake its case—before the American public as environmental and energy security concerns again take center stage.

There is little argument on the contribution of nuclear power to U.S. energy security. In terms of its abundance as an energy source, uranium, when used in breeder reactors, is much more plentiful than coal, oil, or gas. Utilities depending on oil and gas must compete with transportation, home heating, process heat, and consumer products sectors of the economy. There is no significant competing use of uranium, however. As nuclear advocates are fond of saying, it would take the combined coal output of the United States and USSR to displace the current nuclear generated electric power in the United States.

Nuclear provides about 17% of U.S. electric power capacity. If all plants currently under construction are completed by the mid-1990s, that share would grow to 20%. In terms of production, nuclear already is near that one-fifth market share.

But that is a big *if*. There has been effectively a de facto moratorium on U.S. nuclear plant construction, since no new plants have been ordered since 1979.

Anti-nuclear activism, born in the 1960s and growing in the 1970s, took on a grassroots Nimby profile with the accident at the Three Mile Island nuclear power plant in 1979. Although it resulted in no injuries, deaths, or likely long-term health effects—the chief damage being to the utility's investors—it was a crippling body blow to the nuclear power industry in the United States. Perhaps as damaging to the nuclear cause as TMI was the coincidental release of the film "The China Syndrome" and the later release of the film "Silkwood," both of which served to dramatize, and thus exploit, the American public's fear over nuclear power that had already been a residual legacy of the Cold War.

Consequently, nuclear had been effectively institutionalized in the American consciousness as a bogeyman before the horrific accident in 1986 at Chernobyl in the USSR. Chernobyl gave antinuclear activists the trump card they needed. The likelihood of a Chernobyl-type accident occurring in the United States is infinitesimally small. However, the Soviet disaster, with its grim death toll and prospects for long-term damage to the population's health with the massive release of radiation that spread around much of the world, served to validate the antinuke groups' claims that nuclear power was too deadly to be permitted, especially in *my back yard.*

The calcification of the nuclear power industry begun by concerns over public safety and health hardened further as nuclear plant construction and operating costs skyrocketed in the 1970s and 1980s. Delays and cost overruns mounted as a result of mandated design changes requiring added multiple and redundant safety equipment, long delays in construction, and sometimes just poor management.

Nuclear advocates have long touted the relative health, safety, and economic benefits of nuclear power. Risk experts find greater health and accident hazards associated with coal-fired plants than with nuclear plants. Evidence shows that nuclear power plants in the United States that have not suffered from lengthy construction delays can compete with coal-fired plants. The pro-nuke lobby points to France and West Germany as examples of nations with access to plentiful reserves of low-cost coal using nuclear safely and effectively to generate electricity.

Safe and economic may be the reality, but it is not the perception of nuclear power in the United States. If the nuclear industry can ever overcome the public perception of the former, the latter is likely to follow.

Ironically, the same factors that have all but buried the nuclear power industry in the United States—environmental, safety, and health concerns—may have breathed new life in pro-nuke forces.

The environmentalists seemed in 1988–89 to have painted themselves into a corner on nuclear energy. If oil, coal, and gas contribute to the greenhouse effect, acid rain, and smog, alternative fuels are not commercially available on a widespread basis, and conservation imposes economic burdens, then what is left?

There was a startling turnabout for some environmentalists and "green-oriented" politicians in the devastating heat and drought of summer 1988. The 48-nation conference on global warming in Toronto that summer called for "revisiting the nuclear option—through improved engineering designs and institutional arrangements, nuclear power could have a role to play in lowering CO_2 emissions."

Sen. Timothy Wirth (D.-Colo.), a favorite of the environmental lobby, called for renewed research on "safe" nuclear reactors because of air quality concerns. In 1989, Wirth reintroduced into Congress the National Energy Policy Act of 1989 aimed primarily at mitigating the greenhouse effect. It would cut U.S. emissions of carbon dioxide by 20% by the year 2000, authorize $600 million in fiscal 1991–93 to develop renewable energy sources, and redirect $500 million of DOE funding for three years for research and development on advanced nuclear reactor designs. Separately, Sen. Bennett Johnston (D.-La.) introduced legislation providing for a prototype advanced reactor design.

Even Stewart Brand, millionaire philanthropist and founder of the bible of the environmental movement in the 1960s and 1970s, the *Whole Earth Catalog,* predicted that most environmentalists would be pushing nuclear power plants in the 1990s.

This turnabout created a significant rift in the environmental lobby between the pragmatics seeking viable alternatives to fossil fuels, and the utopians insisting that only "soft" energy paths are acceptable.

The pro-nuke forces have been quick to exploit this window of opportunity to press their case.

Early in 1989, the U.S. Council for Energy Awareness noted that the revolution in nuclear safety practices spawned by Three Mile Island is paying substantial current dividends. It cited an industry survey showing that utilities have made progress in almost every category of nuclear plant performance—operating not only more safely, but more efficiently.

Among those improvements:

- The number of unplanned automatic scrams (shutdowns) among U.S. nuclear plants plunged 72% to 2.1 in 1988 versus 7.4 in 1980.
- Operating gross heat rates are lower, a sign of better performance and maintenance.
- Worker exposures to radiation fell by 61% in 1988 from levels in 1980.
- The volume of low-level solid radioactive wastes has been cut by 72% during 1980–88.

Although no new nuclear plants have been ordered since 1979, the industry has shown how much it is capable of contributing to the nation's energy supply in that time.

In 1979, 71 licensed nuclear units generated 255 billion kw-h in the United States. In 1988, 111 units generated 522 billion kw-h. In 1989, nuclear commanded 19.6% of U.S. electricity compared with 11.4% in 1979. By 1989, nuclear power output amounted to double that of hydroelectric power and almost one-third more than the combined output of oil and gas plants.

Bear in mind, however, that added contribution came from plants that were already on order or under construction more than a decade ago.

The nuclear industry in 1989 projected that, assuming electric power demand grows 2–3%/yr through the turn of the century, the United States will need 120,000–220,000 MW of new capacity by the year 2000 beyond what was then under construction. Can nuclear meet that need?

The industry seems to think so. Westinghouse Electric Corp. in late 1989 said that for the first time in a decade, nuclear energy again is a real option. In addressing the Nuclear Energy Forum in San Francisco, Westinghouse Vice-Pres. Richard Slember cited utility concerns about the environment, new plant designs, and institutional arrangements intended to reduce financial risks to utilities as again making nuclear energy an appropriate technology in the United States.

What is still needed to decrease financial risks of nuclear plant construction, said Slember, is that future plants would use simpler, pre-approved designs, be built on preapproved sites, and have a construction period of three years. He also foresees new creative financing methods to sharply cut those financial risks, such as turnkey construction, wherein builders bear risks instead of utilities; equity sharing of construction projects; construction of nuclear plants by independent generating companies; and government participation in a "lead" project.

But there is only a short window of opportunity, Slember warns. His contention is that the nuclear industry, with more than a decade of no new plant orders, is in danger of losing the institutional memory and scientific and engineering talent needed to build new plants in the United States.

What's needed, say nuclear officials, is a push by the federal government to modernize the nuclear plant licensing system, reform the regulatory process, and proceed aggressively with plans to select a high level nuclear waste site in Nevada. Also needed are continuing assistance by DOE to develop and certify advanced, standardized reactor designs and prior agreement between utilities and state regulatory commissions on the need for new electric capacity and conditions for equitable rate treatment.

All of this begs a point, however. However much the nuclear option might alleviate the somewhat vague, fuzzy fears over global warming from the greenhouse effect, can it overcome the deeply entrenched public fear of nuclear power and the concomitant powerful Nimby syndrome?

It certainly can't be lost on the public that the most daunting challenge facing DOE's new secretary, Adm. James Watkins, is the enormous task of cleaning and restoring nuclear weapons waste sites. The cleanup and restoration program for the Hanford, Washington, weapons site will cost as much as $1 billion/yr. Multiply that by the dozens of nuclear arms waste sites around the country, and the United States could face staggering cleanup costs in the 1990s.

Dubbed the flagship of the nuclear waste cleanup effort, the Hanford program is likely to undergo intense scrutiny. Every little imbroglio, each setback, each cost overrun, will be pounced upon by the Nimbys and the anti-nukes and dutifully reported in grisly detail on the evening news. It is likely to further prejudice the case against nuclear power.

A reigning symbol of the anti-nuke forces in the United States has been the Shoreham, Long Island, plant. The last major unit finally obtained a license in early 1990. But for every Shoreham finally surviving the process, there are several like Black Fox in Oklahoma—killed in its inception—and Fort St. Vrain, near Platteville, Colorado, scheduled for decommissioning in June 1990. Fort St. Vrain, the nation's only high-temperature, gas-cooled commercial reactor, suffered operating difficulties from the beginning.

Public Service Co. of Colorado in late 1988 accelerated plans to halt nuclear operations at the plant and launched a study of converting the plant to a fossil-fueled plant.

Such a conversion already is under way at Midland, Michigan, where a consortium of companies have completed the process of converting a mothballed nuclear power plant to a gas-fired cogeneration plant. The companies involved are looking at further opportunities for nuclear-to-gas conversions.

Will concerns over global warming and acid rain keep the United States from following the example of Sweden, where the parliament in 1988 endorsed the first step to eliminate all nuclear power by the year 2010?

The nuclear program in Sweden has been a technical and economic success, providing almost half of the nation's electricity. But the shadow of Chernobyl has cast a pall over nuclear power in that nation. So with nuclear fading, how will Sweden meet rising demand for electricity? It has a sizable wind power program, and more than half of its electricity comes from hydro. But wind power's cost at best will be four times that of conventional electric power costs in Sweden. And hydro could grow by another third, but Sweden has banned further damming by protecting remaining rivers. Coal would have to be imported, and new coal-fired plants are expected to produce electric power at twice the cost of current electric power, not to mention contribute to acid rain and global warming. The solution? Perhaps it is natural gas, with its relatively low cost, imported from fields off Norway and relatively benign environmental effects. Sweden and other Scandinavian countries already are scrambling to line up North Sea and Soviet gas supplies.

In the end, despite some wobbling among environmentalists on whether nuclear is a lesser evil than once perceived, the environmental lobby still is largely set against it. Early in 1990, a report by Ralph Nader's Public Citizens Organization on behalf of several dozen environmental and anti-nuclear groups in the United States came to the conclusion that the nuclear industry has not proved that it can stand on its own economically, or deal with the problems of safety and water disposal.

"The nuclear industry is now seeking to exploit (concerns over U.S. energy shortfalls and the environment) by portraying itself as the solution to the nation's energy and environmental woes," the report said. Despite the promises of nuclear power, it said, the industry has grown more costly, safety has not been improved, and the problem of radioactive waste has not been solved.

If there were widespread public support and broad acceptance by the environmental community for nuclear power, then the nuclear industry might have a fighting chance to exploit that window of opportunity. Without that, nuclear's contribution to the surging electric power demand growth in the 1990s is likely to be marginal.

Questions about Coal

• •

For all the talk about environmental concerns and coal's "dirty reputation," it might be surprising to realize that coal has enjoyed a significant comeback in recent years.

During most of the 1980s, coal producers endured soft markets, in large part because of the lingering effects that the recession of the early part of the decade had on electric power demand. That turned around in 1988, because of the burst in demand for electric power caused by the record heat and drought of that summer. The same heat that drove electric power demand to record highs brought with it drought that squeezed hydro's capability to meet that electric demand growth. The National Coal Association estimated U.S. coal consumption at a record 958 million tons in 1988, up from 925 million tons in 1987 and 893 million tons in 1986. NCA projected a further increase in 1989 to 966 million tons.

Most coal producers project that demand for U.S. coal in the 1990s and beyond will far outstrip demand seen in the 1980s. That will be due in large part to sharply higher electricity demand growth in the United States, a spurt in demand for steam coal in the Pacific Rim, and higher demand for steam and metallurgical coal in Europe resulting from decreased production from high-cost mines there.

The strongest demand growth for coal will occur among low-sulfur, high-quality coals. Any new acid rain or greenhouse legislation is likely to spur that growth at least in the interim, as utilities are obliged to back out cheaper if more polluting grades of coal. U.S. demand for low-sulfur, high-quality coals from the Rocky Mountain basins is expected to double or perhaps triple in the 1990s.

Coal's dynamic comeback would have been even more dramatic if the nuclear "bubble" of ordered capacity had not been a factor. As nuclear continues to suffer in doldrums, and electric demand growth continues, coal's strength could remain evident through the turn of the century and beyond.

U.S. coal's ability to meet that demand is clear. The resource is there. The nation currently has a reserve life of coal, at current consumption rates, of 200 years. Ultimate potential recoverable coal could have a reserve life of as much as 400 years.

There are warning signs for coal on the horizon, however. The number of utility-owned, coal-fired power plants entering commercial service hit a record low in 1988, according to the Utility Data Institute. The number of new coal plants that year was only two, the smallest number since utilities began building coal-fired power plants at the turn of the century. That compares with 22 new coal-fired units entering service in 1969, the year the National Environmental Policy Act was passed. The number of new coal plants added to service had remained at more than 20 units/yr until 1983, when it fell to 14.

So the critical questions are: How much will new clean air legislation affect demand for coal in the long term as well as the short term; how will coal meet the challenge under tighter standards through new technologies or new applications of existing technology; and will those factors affect the economics of America's most abundant, if "dirtiest," energy resource?

The approval of amendments to the CAA—likely to occur in 1990—almost certainly will crimp the growth of coal in the U.S. electric power generation market in the long term, and perhaps in the short term, depending upon how stringent the final acid rain standards will be.

The Bush administration's clean air bill calls for a 10 million-ton reduction in sulfur dioxide emissions by 2000, of which 90% would come from utilities.

The Bush bill would set a cap on sulfur dioxide emissions, beginning in the year 2000, that would be based on 1985–87 capacity utilization and an emission rate of 1.2 lb/million BTUs. It also set the year 2003 as the deadline for clean coal technologies to be in place, although many scientists think that won't happen, conservatively, before the year 2005. At the same time, the Bush bill inhibits the demonstration and commercial availability of the complete range of clean coal technologies. Further, the administration bill would limit the use of coal by allowing governors to prohibit the use of out-of-state coal.

The Bush bill is considered to be more a starting point in CAA reauthorization than a blueprint for final action. The environmental lobby considers the administration clean air initiative as not stringent enough, while industry worries about the economic impacts of the bill's provisions. A final compromise is likely to result in a strong emphasis on utilities' freedom of choice by dictating emission standards and goals instead of dictating specific technologies, with some limited calls for mandated scrubbing equipment.

The window is likely to be a short one. Assuming acid rain legislation is enacted in 1990, EPA will need about a year to draw up regulations for states to use in developing implementation plans, according to Burns & McConnell. Allowing a year to develop the states' plans and another year for EPA approval would bring the industry to 1994.

"Therefore, only about two years would be available for affected utilities to finalize and implement first phase compliance measures expected to be required by 1996," Burns & McConnell said. "That's not much time if any new construction is necessary."

Industry generally would like to see any acid rain laws focus on goals, not specific technologies, with Congress and the White House limiting directives to emissions reduction requirements and allowing scientists and businesses to achieve those goals. Flexibility encourages innovative and cost-effective approaches, and thus ensures a greater level of compliance.

The ultimate tone of acid rain legislation could rest on the final results of a massive study authorized by Congress. The National Acid Precipitation Assessment Program (NAPAP) is a 10-year, $500-million research effort funded by the federal government. It was expected to be completed in 1990. Interim results from the study suggested that there will be no easy, clear-cut solutions.

NAPAP's preliminary results showed that damage to the environment from acid rain is not as diverse or as extensive as previously thought. Furthermore, acid rain levels in rainfall have not changed appreciably despite cuts of 23% in sulfur dioxide emissions since 1973 and 14% in nitrogen oxides emissions since 1978—at the same time coal use jumped 45% during 1973–88. NAPAP essentially concluded that society has time to continue research on acid rain and its effects, along with development of new emissions control technologies.

In the meantime, how to control those emissions that contribute to acid rain is the heart of the issue's controversy.

Utilities want to avoid a massive program of adding scrubbing equipment to clean up sulfur dioxide and nitrogen oxide emissions. More than 160 scrubbers have been installed at power plants in the United States, and another 20 are planned or under construction.

Although they meet the air quality standards called for under the CAA, scrubbers carry their own huge costs, not to mention added environmental and energy efficiency burdens. Some of the big scrubbers can cost more than $100 million and eat as much as one-third of the total electrical output of a power plant.

According to Utilicorp United, Kansas City, Missouri, installing scrubbers at each of its coal-fired power plants would cost $4 billion during 10 years. Further, conventional wet scrubbers produce big volumes of sludge that must be disposed of properly in appropriate landfills. And the wet scrubbers only take care of sulfur dioxide, not nitrogen oxides or carbon dioxide. In fact, adding scrubbing equipment forces a power plant to burn more coal, thus emitting more carbon dioxide—of greenhouse fame—than before.

Nevertheless, power plants are expected to spend $13 billion for scrubbers in the 1990s, according to a forecast in mid-1989 by McIlvaine Co., Northbrook, Illinois. If all new power generation needs were to be met by coal-fired power plants using scrubbers, scrubber purchases would run $3 billion/yr, according to McIlvaine.

Pretreating coal or switching to low-sulfur coals creates new problems. Switching to low-sulfur coal involves lining up a new supply source and perhaps coal washing, not to mention the new fuel's effects on boiler and other plant operations.

For utilities, the preferred alternative to scrubbing equipment is an accelerated program of clean coal technology development. Clean coal is evolving into a healthy cottage industry funded by the Electric Power Research Institute and a $5-billion, five-year research and demonstration program funded by the DOE. Utilities maintain that, over the long term—say, the year 2010—clean coal technologies will produce an equal or greater amount of emissions reductions at a cost of about 20% less than that for scrubbers under proposed acid rain legislation.

Repowering with new, cleaner coal-based technologies offers great promise. The hitch is in timing. In the near term, the emphasis on clean coal approaches likely will be on scrubbing, pressurized fluidized bed combustion, atmospheric fluidized bed combustion, hot gas cleanup, direct coal liquefaction, and underground coal gasification.

In the longer term, the emphasis will shift to surface coal gasification, advanced combustors, fuel cells, and indirect coal liquefaction.

By early 1990, DOE had chosen 40 projects under three rounds of solicitation for clean coal technology demonstration projects. Under the program, DOE plans by 1992 to help finance 50–75 demonstration projects. Early in the program DOE was blistered by environmentalists for weighting the program heavily with repowering projects at the expense of retrofit/scrubber technologies. However, the last round, at year-end 1989, had more than half the selections in the retrofit camp. Environmentalists want the mandatory installation of scrubbers to achieve emissions reductions more quickly.

The question of timing is critical to the controversy. The most appealing technologies, fluidized bed combustion and coal gasification, may not be widely available on a commercial basis until the latter part of the 1990s. Environmentalists insist that the future costs associated with delaying repair of environmental damage and health care costs will be far greater than the short-term costs of scrubbers. Utilities contend that the situation for acid rain has stabilized and that there is time until the new repowering technologies and reduced cost scrubbing approaches are commercialized. Again, the widely divergent views probably will lead to a messy compromise under CAA amendments, leaving neither side especially satisfied.

However, there appears to be an interim solution that could form a happy compromise and provide a wonderful marketing opportunity for the gas industry: cofiring.

Cofiring, or the burning of limited volumes of natural gas in combination with coal or fuel oil in utility and industrial boilers, reduces emissions, improves operations, and boosts economic performance. Plants can operate at full capacity and stay within air quality standards. At the same time, cofiring enables a plant operator to switch to lower quality coals, rein downtime, improve combustion efficiency, and cut maintenance of downstream equipment.

The potential for cofiring can be significant. The pioneer in the field, Consolidated Natural Gas Co., estimates cofiring will mean an additional 40–45 billion cu ft/yr of new business for the company by the end of 1992. CNG's approach calls for convincing a plant operator of the economic and technological benefits of cofiring and then finding an easy way to introduce gas into the plant, such as igniting the coal, followed by a gradual buildup in the gas feed to as much as 20% of the total fuel load.

The environmental results can be dramatic. A demonstration project in Pittsburgh in 1989 sponsored by the Gas Research Institute and Duquesne Light

Co. produced reductions in sulfur dioxide emissions of 12% and nitrogen oxide of 25% by burning as little as 4–9% with the coal.

A newer cofiring process, reburn, goes much further, using the chemical properties of natural gas to cut nitrogen oxide emissions by as much as 60%. There are technologies involving coal/natural gas combustion that could result in reductions of as much as 95% of both key acid rain pollutants.

Getting access to gas won't prove a problem for most power producers that need it for cofiring. Interstate Natural Gas Association of America (INGAA) notes that the top 100 emitters of sulfur dioxide in the United States are an average 5 mi from one or more gas pipelines and that 13 have natural gas hookups.

Cofiring is just one example of where the natural gas industry can make significant headway in securing more of the surging electric power market. Other approaches include repowering boilers to burn natural gas, switching to gas on a seasonal basis, and installing gas-fired combustion turbines.

Altogether, additional demand for gas created by electricity consumption growth could total as much as 2 TCF/yr—based on new generating capacity already on the drawing boards, according to American Gas Association (AGA).

Gas advantages over coal and oil are manifold in power plant use. It takes about three to four years to site and build a gas-fired power plant. A coal plant takes about four to seven years.

Gas is the winner in operating economics as well. Figuring in operating costs and dispatch reliability, a gas-fired, combined-cycle power plant is more economic than an oil-fired plant even if the delivered price of the gas is as much as 20% higher than fuel oil. The payout for an IPP or cogeneration power plant can be as short as five years for one that is gas-fired compared with one that is coal-fired. Capital costs for gas turbine-generated power ranges $400–700/kw. For a coal-fired facility, the cost jumps to as much as $1,400/kw. Longer term, however, the comparisons of economics tend to even out. A 50-year coal contract for an IPP can be had for as little as $1.50/MMBTU. Gas contracts usually do not run for more than five years, owing to the historic volatility of gas prices.

That leads to the biggest single stumbling block for the gas industry in making significant inroads in the electric power market in the 1990s: developing supply contracts for 15–20 years that are acceptable to gas producers and power plant operators alike.

Power plant operators generally want a gas price that is competitive with coal, because they base the price escalators in their contracts with IPPs on the price of coal. Gas producers, however, want escalators tied to finding costs or indexed to fuel oil or an alternate gas supply. In some areas of the country, gas producers will be hard pressed to meet long-term contracts. In the Gulf of Mexico, for instance, fields tend to play out in as little as five years. So to fulfill a long-term contract, a gulf producer will have to assume that he can replace his initial reserves three more times at a competitive cost near the same pipeline as under the original contract.

The disappearing gas "bubble," or surplus deliverability, is bringing that concern home to power producers. Many in the industry surmise that the gas bubble is already effectively gone, as witnessed by the supply curtailments in the past two winters in California and elsewhere in the nation.

It is not a question of the potential resource. Although the reserve life of U.S. gas totals more than 20 years at the current rate of consumption, spiraling demand is shrinking that number.

In a 1989 study, the U.S. EIA projected that the United States may be using twice as much natural gas to generate electricity by the year 2000 as it does now (Figs. 6–1 and 6–2). EIA sees the U.S. gas resource as adequate to meet that extra demand. Assuming an economic life of power plants of about 30 years, natural gas use to generate electricity will jump from 3.2 TCF/yr to about 6.2 TCF/yr in the year 2000 and beyond, as the gas-generating plants are retired by the year 2030.

However, since it is much cheaper to completely revamp an existing gas plant than it is to build any kind of new grassroots plant, utilities may opt to go that route instead. If that happens, according to EIA, the life cycle of power plants will be greatly extended, perhaps even doubled, eliminating a post-2000 slide in gas demand by the electric power industry.

"This implies a continuation of the 6.2 trillion cubic feet annual gas demand to the year 2030," EIA said. "The resource base now generally believed to exist in the United States would not be able to handle such a massive demand for so long."

The United States will have used up about 80% of its natural gas resources by the year 2030, EIA estimated (Fig. 6–3). EIA pegged the U.S. gas resource in 2030 at about 180 TCF, compared with 830 TCF in 1988. By 2030, the nation will have to depend on imports for about 26% of its gas supply needs, from about 7% in 1988. In addition, about 10% of all pipeline quality gaseous

NATURAL GAS: FUEL OF THE FUTURE

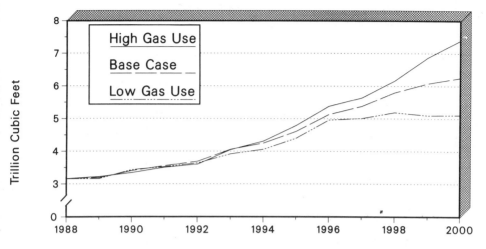

Note: Includes natural gas consumed by electric utilities and nonutilities (industrial facilities and others).
Source: Energy Information Administration, Office of Coal, Nuclear, Electric and Alternate Fuels, Low-Gas-Use Case --IGHILG.D1206883, Base Case--IGMD89.D1205885, High-Gas-Use Case --IGL0HG.D1206884.

Figure 6–1 Natural gas consumed to produce electricity for 1988–2000.

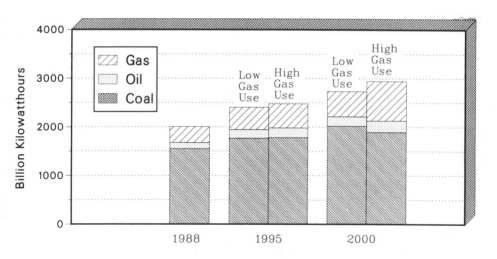

Note: Includes fossil fuels consumed by electric utilities and nonutilities (industrial facilities and others).
Source: Energy Information Administration, Office of Coal, Nuclear, Electric and Alternate Fuels, Low-Gas-Use Case--IGHILG.D1206883, High-Gas-Use Case--IGL0HG.D1206884.

Figure 6–2 Electricity generation by fossil fuels for 1988, 1995, and 2000.

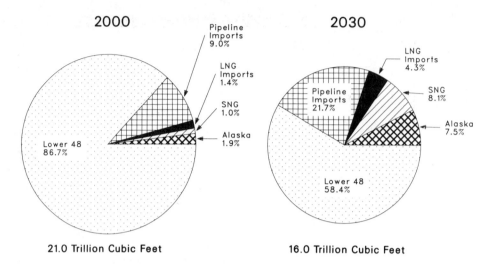

2000

2030

21.0 Trillion Cubic Feet

16.0 Trillion Cubic Feet

Note: SNG is Synthetic Natural Gas. LNG is Liquefied Natural Gas.
Sources: Energy Information Administration, 2000 projections: Base Case--IGMD89.D1205885, 2030 Estimates: Office of Oil and Gas.

Figure 6–3 Natural gas supply, 2000 and 2030.

fuel used in the United States in 2030 is likely to be synthetic, produced mainly from coal and liquid hydrocarbons, EIA projected.

Because of the tightening supply/demand balance for natural gas, the average wellhead price for gas is likely to double (in real dollars) by the year 2000. Consequently, utilities are likely to retire most gas-burning plants at the end of their economic lives, without any effort to extend their lives or convert them to other, more economic fuels, EIA said.

The gas industry was quick to dispute EIA's conclusions, citing a study a year earlier by DOE to assess the potential for U.S. natural gas. The DOE study concluded that the recoverable resources of conventional gas totaled 1.059 quadrillion cu ft in the Lower 48 states. Of that total, 800 TCF lies in conventional reservoirs and can be extracted at a cost of less than $3/MMBTU. The rest is economic at $3–$5/MMBTU.

There is also the massive Alaskan North Slope gas resource to consider, at more than 30 TCF. Combined with frontier gas supplies in Canada, the North American Arctic's gas reserve potential is more than 60 TCF. Of course, price is the key. It is not likely that arctic gas supplies will be economic until the turn of the century.

In the meantime, there are two other sources of potential gas supplies to help bridge the gap to the 21st Century: coalbed methane and LNG.

Coalbed methane proved a surprise to the industry in 1988–89, generating frenzied drilling activity in areas like Alabama's Black Warrior basin and the San Juan basin of Colorado and New Mexico. Relatively low costs, aided by a federal tax credit scheduled to expire in 1991, have led to a significant addition to the U.S. gas resource base. Recognition of this incremental increase is likely to result in the tax credit being extended through the 1990s.

Worldwide, the demand for LNG is accelerating as the economics of gas improve and more nations—notably those crude exporters seeking additional markets for their hydrocarbon resources—are developing low-cost sources of gas for LNG projects. Countries like Algeria—previously a major exporter of LNG to the United States—are selling LNG at scarcely breakeven prices to ensure contracts that will certainly look better in the future. As a result, the United States again is importing substantial volumes of LNG after a period of negligible LNG imports in the latter 1980s. Joining Algeria in the 1990s in competition for the U.S. LNG import market will be Norway, Nigeria, Qatar, Venezuela, Indonesia, Australia, and Papua New Guinea.

In short, the gas supplies are likely to be available to meet expected demand in the electric power sector in the early part of the next century. Will those future incremental supplies be cost-competitive?

There are three factors to consider here. One is the uncertainty over future costs of generating power with coal. If scrubbing is mandated and not directly subsidized, gas almost certainly will boost its market share in power generation at the expense of coal in the 1990s. Studies show gas at even a sharply higher price can compete effectively with scrubber/coal power generation and provide more environmental benefits.

Secondly, there is the near-term, price-depressing effect of dual fuel boiler capability. Almost 4.9 TCF of gas demand in the combined power plant and industrial sectors is burned in such dual-capable facilities. The switches can be made in hours or even minutes and often on price differences of a few cents per million BTUs. About one-third of total gas consumption is subject to fuel switching—about 95% in power generation and 30% in industrial plants.

Gas now must compete with residual fuel oil to avoid big market losses. So it follows that as oil prices rise in the 1990s—most likely modest increases averaged out over the decade—gas prices will keep pace, rising enough to stimulate drilling from current depressed levels but not so sharply as to switch to fuel oil or improve the relative economics of scrubbed coal. As drilling accelerates, the resource base in the United States, Canada, and overseas also will expand. This follows the same pattern established for oil in the late 1970s and early 1980s, and there is no reason to doubt that gas will do the same. Of course, natural gas could run into the same fate that oil did in the 1980s: too much drilling for any kind of prospect inflating its cost, and a skyrocketing price adding too much supply at the same time it spurs conservation and fuel switching.

There are critical differences for natural gas, however. Unlike with oil, a reasonable cost resource of natural gas has yet to be fully tapped—and there are the natural price constraints of a deregulated market. The United States just does not face the same kind of energy security or trade balance threat with gas that it does now with oil.

AGA contends that the U.S. surplus of gas deliverability is gone and that there will come several years of balance in U.S. gas supply and demand.

AGA figures that incremental supplies of as much as 1.35–2.1 TCF/yr of gas could be made available within three years if needed (Table 6–1).

Table 6–1 U.S. Incremental Gas Supply Potential

	Within One Year	Within Three Years
	——BCF——	
Nonproducing reserves	250–750	500–1,000
Canadian gas*	400	500
Accelerated infill drilling	50–100	100–400
LNG**	40–100	200
Total	**740–1,350**	**1,300–2,100**

* Volumes are in addition to 1.283 TCF of imports estimated in 1988.

** Volumes are well below the physical limits of existing terminals.
Source: American Gas Association

Gas well completions are rising as well. AGA estimated U.S. producers would complete 10,040 gas wells in 1990, compared with 8,840 in 1989 and less than 8,000/yr in 1986–88.

The winter of 1989–90 gave the gas industry a healthy taste of just how tight the situation can get. A sudden, severe, and widespread cold snap spiked demand for gas and froze off significant production volumes when wells froze up. That shot the spot price of gas up to levels not seen in almost five years.

Those are signs of health that is certain to turn more robust in the next few years. The gas industry's long-awaited cure will come in large part because of the environmental ethos that is seeping into the national consciousness and permeating every part of the American fabric. It is abundantly clear that natural gas is the environmentally preferred fuel of the 1990s.

References

1. Arthur Andersen & Co./Cambridge Energy Research Associates Inc., *Electric Power Trends,* 1989.
2. Russell Leavitt, Dr. Mark Luftig, Salomon Brothers Research Department Analysis, April 1989.
3. Investor Responsibility Research Center, *Mergers and Financial Restructuring in the Electric Power Industry,* May 1988.
4. Burns & McDonnell, *Future Power Generation,* Winter 1989–90 (Consultants' report).

7

STRATEGIES FOR U.S. EXPLORATIONISTS

The strategy of U.S. petroleum explorationists in the 1990s can be summed up in one word: *access.*

All other environmental issues regarding U.S. oil and gas exploration flow from that. Without the prospects, the U.S. petroleum industry stands virtually no chance of replacing production and building the reserves base for America's future energy needs.

There have been major advances in technology in the 1980s to improve the petroleum industry's ability to find and produce more oil and gas. Three-dimensional seismic interpretation allows better identification of potential hydrocarbon-bearing structures and improved detailing of existing reservoirs. Horizontal drilling has proven to be an exciting evolutionary advance, providing a cost-effective way to bring in a more productive well by exposing more of the reservoir's pay zone to a wellbore. Enhanced recovery techniques such as steamflooding and carbon dioxide injection have sharply boosted production and increased ultimate recoverable reserves in mature fields—an enhanced oil recovery program at Prudhoe Bay added at least two years to the peak production level of North America's biggest field.

But even the greatest advances in technology have their economic and technical limits. U.S. oil and gas companies must have access to untapped lands, especially federally owned lands, in order to replace their reserves, or they face self-liquidation. It is simple economics: A producer of a nonrenewable commodity must continually find new sources of supply as his existing source is being exhausted, seek to import it, or get out of the business of supplying that commodity.

The staggering collapse in the U.S. oil and gas potential resource base in the 1980s can be seen in one telling statistic: According to Minerals Management Service, the mean, undiscovered recoverable U.S. petroleum resource has plummeted to 49 billion bbl of crude oil and 399 TCF of natural gas in 1987 from 83 billion bbl of crude and 594 TCF of gas in 1980. Bear in mind that those are postulated numbers, not an estimate of the actual resource. A major part of the decline is because of disappointing results in some of the once-promising unexplored regions, such as the Georges Bank off the U.S. East Coast and the highly prospective basins of the Bering Sea off western Alaska. Even at that, the reduced estimate is roughly double that of current U.S. oil and gas reserves.

Production in the United States plunged by about 500,000 B/D in 1989, largely the result of the 1986 oil price collapse that decimated the U.S. oil industry and its high-cost, low-production wells and sparked U.S. oil demand to its highest levels in history. The prospects are even worse for the future. The discovery and development of oil at Prudhoe Bay, Kuparuk River, and other fields on Alaska's North Slope forestalled the United States' being dependent on imports for more than 50% of its oil needs during the energy crises of the 1970s and 1980s. But Prudhoe led a North Slope production decline beginning in 1989 that will slice output from North America's premier petroleum province to less than 500,000 B/D by 2000 from almost 2 million B/D at the peak in 1987–88. The Alaskan North Slope has provided the United States with as much as 25% of its oil supply in the past 12 years.

Meanwhile, in July 1989, the level of petroleum imports into the United States topped the 50% mark for the first time since 1977. Imports reached 8.552 million B/D, or 50.4% of U.S. deliveries totaling 16.98 million B/D. That compares with 31.5% in 1985. U.S. oil imports climbed still further in 1990, reaching 52% of U.S. supply at 8.511 million B/D in May 1990, up 11.7% from the previous year.

According to the American Petroleum Institute, increases in oil consumption accounted for about two-thirds of the growth in imports in recent years, and declining production mostly accounted for the balance. The situation changed in 1989, when the reversal in the direction of Alaska's production profile caused declining production to be the single biggest factor in the alarming rise in U.S. oil imports, and U.S. consumption slowed slightly.

For the near term, at least, that trend appears to be irreversible. Estimates of future U.S. dependence on oil imports range to as much as 75% in the

1990s, even if consumption remains flat. At the same time, the productive capacity of crude exporting nations outside the Organization of Petroleum Exporting Countries (OPEC) will fall in the next few years as oil demand continues to climb. It is just a matter of a few years before OPEC controls enough of the world's oil supply to again be able to dictate price levels pretty much wherever the member nations choose.

The OPEC nations have expressed that it is in their interest to avoid a repeat of the price shocks of the 1970s—which spurred conservation and the rise of non-OPEC production—or the 1980s—which all but bankrupted some of those countries. However, a betting man would have takers standing in line if he were to declare that there would be no more oil supply disruptions and concomitant price spikes, that Israel and its Arab neighbors would make peace, or that Islamic fundamentalism will not affect moderate Arab states. Not only are there no guarantees against future oil price spikes, the odds are not that good that there won't be.

The environmental groups preach conservation and fuel-switching to offset rising oil demand in the United States. But both carry with them technical and economic limits. With the surging economies of Japan and western Europe, not to mention the new competition from South Korea, Taiwan, and Singapore (or possibly a reborn eastern Europe later in the decade), the global marketplace for the United States will be a much tougher arena in the 1990s than it has been. At the same time, a nation already beset with huge budget and trade deficits will be called upon to underwrite the infant democracies that are popping up around the world seemingly with each new month. With already staggering social problems and apparently intractable political aversion to taxes, the competition for American spending in the 1990s will be fiercer than imaginable now.

Will the American public, for all its strong support (at least when asked by poll takers) for environmental protection, tolerate a second-class economy marked by massive unemployment and the kind of stagflation that drains the funds available for solving other pressing social ills? Would that not be the price when Japan and a reunified Germany feast on a banquet of low-cost oil from the Middle East as the United States adopts a draconian lifestyle based on conservation and renewable fuels economic only at an equivalent price of $40–$50/bbl of oil?

The likelihood is much greater that the United States also will share in that feast of relatively low-cost oil—at a price of 75% import dependency—and perhaps at a much higher oil price than most experts expect right now. The experts were resoundingly, almost unanimously wrong about energy supply, demand, and price patterns in the 1970s and early 1980s. Why should this decade prove different?

DOE projections show that, although energy efficiency in the United States will continue to improve—in large part as a result of the turnover in older, less energy-efficient vehicles and appliances—overall energy use in the United States will continue to climb this century. DOE studies indicate that each BTU will produce 15% more economic output in the year 2000 than it does today. However, because the U.S. gross national product is expected to grow by at least one-third this decade, overall energy use will rise 14% by the year 2000. That translates to an additional 1.6 million B/D of oil by the turn of the century—beyond current record levels of consumption and imports.

To keep energy consumption at existing levels, according to API, oil prices would have to jump to $38/bbl in the year 2000 from $13/bbl in 1987. To reduce consumption by even as little as 10%, oil prices would have to soar to $61/bbl by the year 2000. That's the kind of skyrocketing oil prices that sent a firestorm of inflation ravaging through the world's economies in the 1970s and early 1980s.

Even halting U.S. oil demand growth would only slow U.S. dependence on oil imports. That's because U.S. oil production will continue to decline, according to DOE projections, by as much as 2 million B/D by the year 2000.

Accordingly, it would behoove the United States, still the repository of an oil and gas resource likely sufficient to support more than half of its petroleum needs for the next few decades, to promote the exploration and development of those resources.

But the United States is the oldest and most thoroughly explored and developed region for oil and gas in the world. All of its major producing basins are now essentially in decline. To replace its rapidly depleting oil and gas reserves, the United States must turn to frontier areas. And that means Alaska and the Outer Continental Shelf. The problem is that environmental opposition has effectively stymied exploration of some of the continent's most prospective petroleum provinces for much of the 1970s and 1980s. And the rebirth of militant environmentalism in 1988–90 may foreclose that critical option forever, unless the U.S. petroleum industry is up to the challenge.

ANWR's Coastal Plain: U.S. Oil's Crown Jewel

• •

Tucked above the Brooks Range of northeastern Alaska is ANWR. Congress created a 9 million-ac Arctic National Wildlife Range in 1960. Twenty years later, Congress passed the Alaska National Interest Lands Conservation Act, increasing the range's area to 19 million ac and renaming it the Arctic National Wildlife Refuge. Even then, the petroleum potential of ANWR's Coastal Plain was apparent. Congress ordered the Department of Interior to conduct a study of the Coastal Plain's resources.

Much of ANWR already is classified as wilderness, and therefore off limits to any resource development. Congress has the option to similarily lock up the Coastal Plain by declaring it wilderness.

One attempt to break the ANWR logjam was the effort among several companies to overcome the apparent competitive advantage that Chevron and BP held through their deal with Arctic Slope Regional Corp. to drill the Coastal Plain's sole well. Although Chevron was blocked from further exploration in the refuge until Congress approved leasing, it could have proceeded with more drilling while the prospective ANWR leases were being readied, thus honing its competitive edge still further. Some companies sought similar land swaps and drilling deals with native corporations holding land on the Coastal Plain. Some of their competitors sought to halt such deals. Meanwhile, others sought through congressional allies to block Chevron from further exploration until the first ANWR lease sale could be held. That sort of divisiveness only served to hurt the industry's advocacy of gaining access to the Coastal Plain.

Ironically, Congress seemed to be moving toward marking up some sort of ANWR Coastal Plain leasing bill in the months before the Exxon Valdez spill. The Senate energy committee began markups after conducting hearings on the issue in early March 1989, with Sen. Bennett Johnston (D.-La.) and Sen. James McClure (R.-Id.) filing the bill.

The new bill sought to earmark 40% of all lease revenues for federal conservation programs. Of the remaining ANWR lease revenues, 50% would go to Alaska and 10% to the federal treasury. It required phased lease sales of not more than 300,000 ac each and let the Interior secretary delete environmentally sensitive areas, leaving oil companies to nominate tracts for the sales. The first sale would be issued within 18 months of Interior's issuance of final sale rules, with the second sale to follow within three years and subsequent sales to follow at two-year intervals. The bill also would have imposed some stringent environmental mitigation measures, including a 5¢/bbl fee on Coastal Plain production to build a $50-million fund for eventually restoring the Coastal Plain or other land on the North Slope.

The Bush administration sought changes in the Johnston-McClure bill, notably the earmarking of conservation funds and the limits on sale acreage.

The bill revived the ANWR debate. Sens. Tim Wirth (D.-Colo.) and Dale Bumpers (D.-Ark.) wanted to tack on an amendment requiring the government to draft an energy plan before allowing the ANWR Coastal Plain to be leased, to ascertain whether the United States really needed ANWR production. Sen. Howard Metzenbaum (D.-Oh.) proposed an amendment that would tighten federal vehicle fuel-efficiency standards. He contended that effort would save about the same amount of oil that ANWR was likely to produce. Sen. Mark Hatfield (R.-Or.) offered an amendment that would have a federal task force develop a national energy plan within 15 months of an ANWR leasing bill being enacted. His amendment survived.

One ironic note to the skirmishes shaping over ANWR leasing in early 1989 was the weary acknowledgment of Sen. Johnston that few hearings would be required for his ANWR leasing bill because four committees in the previous Congress had held 23 hearings on the subject. It was becoming pretty familiar territory by then, and the mood within industry was becoming cautiously optimistic that at least some form of ANWR leasing bill, however diluted or pared down, might be passed in the 101st Congress.

A week after the Senate panel began markups on the ANWR leasing legislation, the Exxon Valdez spilled almost 260,000 bbl of North Slope crude into one of the most pristine, teeming-with-wildlife waterways in the world.

The petroleum industry, federal agencies, some congressmen, and even President Bush quickly pointed out that there was no connection between drilling for and producing oil on ANWR's Coastal Plain and a transportation accident, but it was already all over but the shouting.

ANWR Potential

The petroleum potential of the ANWR Coastal Plain was clearly shown in the study Congress ordered in 1980. Completed in 1987, the Department of

Interior study identified a series of geological structures within a portion of the Coastal Plain (the 1.5 million-ac 1002 area, named after the study) that were among the biggest ever seen in North America. According to a 1947 state report, structures along the Marsh Creek anticline could contain an oil resource of 11 billion barrels.

For decades, evidence of hydrocarbons have turned up in the 1002 area as surface seeps. The Coastal Plain lies between the massive complex of oil reserves in the Prudhoe Bay area—only 65 mi to the west—and the cluster of oil and gas fields found in Canada's Beaufort Sea and Mackenzie Delta to the east (Fig. 7–1). In addition, a number of oil discoveries have been made in the Beaufort Sea directly off ANWR's Coastal Plain. And a giant gas and condensate field, Point Thomson, lies between the Coastal Plain's border and Prudhoe Bay.

There is a consensus among explorationists that the source rocks and reservoir rocks that provide a regime for oil and gas production in the Canadian Beaufort/Mackenzie area and Prudhoe Bay regions are both present in the 1002 area. BP Alaska's Roger Herrera, the dean of North Slope explorationists, cited Upper Cretaceous organic shales as being "full of oil" as a postulated source rock and reservoir rock sandstone with about 25% porosity and excellent permeability seen in outcrops and seeps along the Marsh Creek anticline. There are outcrops along the Marsh Creek anticline where, after a strong rain, the crude oil "just washes down in sheets," Herrera said.

Opponents of ANWR leasing contend drilling there is not worth the effort because the 1002 area holds "only" a 19% chance of a commercial success. However, one in 50 is the historical average for Alaska, according to Herrera.

One in five is wildly optimistic, Herrera contends: "Those are the best odds for oil an explorer has ever had."

If that one in five chance happens, Interior estimates a mean of 3 billion bbl of reserves. However, that estimate is probably very conservative, according to the Office of Technology Assessment (OTA). The estimate is based on "economically recoverable volumes of oil," which likely has changed sharply from the 1981 estimates of production costs used in the study. That year, production costs on the North Slope were at their historic peak. Since the price of oil collapsed, tighter reins on costs and a quantum leap in the learning curve—developed mainly at Kuparuk River, Milne Point, and Endicott oil fields near Prudhoe Bay field—ensure that production costs on the ANWR Coastal Plain will be far smaller than those estimated in 1981.

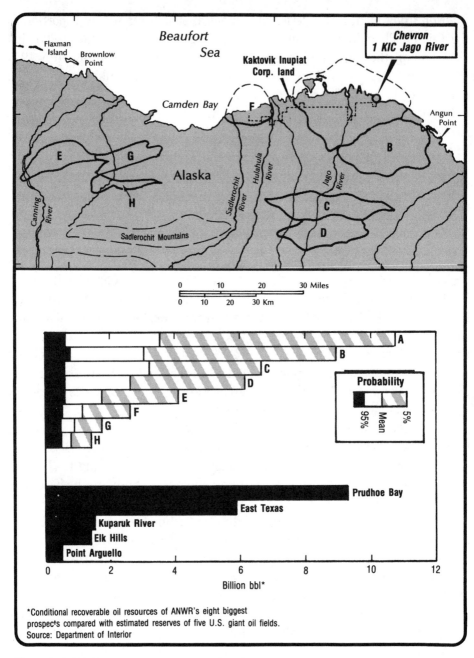

Figure 7–1 ANWR top prospects compared with United States giants.

OTA also took issue with Interior not taking into consideration the possibility that ANWR oil could be developed with two or three moderate-sized fields, even though no single field exceeds the minimum economic size postulated for ANWR. Such smaller prospects were not considered in the study.

Another aspect to ANWR potential that is often overlooked in the debate is the area's potential for natural gas. The Interior study predicted there is a 95% chance of finding at least 11.5 TCF of natural gas in place in the 1002 area. That would not only boost the current North Slope natural gas reserve by about half, it also could render the 5–7 TCF of gas reserves at Point Thomson economic. Should Prudhoe gas be marketed, either to the Lower 48 or exported as LNG to Japan, developing more gas reserves nearby at Point Thomson and in ANWR would make the project economics much more attractive.

Proximity to Prudhoe and TAPS enhances the Coastal Plain's attractiveness to an explorationist even more. The infrastructure already is in place. In contrast, the other remaining giant prospect areas of Alaska, such as the Beaufort Sea, are in relatively deep arctic waters and call for enormous costs and technological risks. There are 300-million-barrel oil fields in the Beaufort Sea that are uneconomic at current oil prices, even in relatively shallow waters. A field only 2 mi from shore, Endicott is marginally economic at about 375 million bbl of reserves but is producing. Another North Slope field, Milne Point, contains only about 60 million bbl of oil reserves, yet it is marginally economic at today's prices because it is literally in the shadow of Prudhoe's infrastructure. Extending that infrastructure to the east could boost the economics of perhaps several billion barrels of oil resources—notably at the turn of the century, when oil prices are likely to be as much as 50% higher than today—on the North Slope and in the eastern Beaufort Sea to the point of commerciality.

There already has been a well drilled on the ANWR Coastal Plain. Chevron drilled the 1 KIC Jago River on native lands in the eastern portion of the 1002 area. It was permitted through an exchange of land that enabled the Kaktovik Inupiat Corp., a native corporation, to let Chevron drill a wildcat on their acreage without running afoul of the congressional ban on drilling in ANWR. The well was drilled to 14,000 ft in the winters of 1984–85 and 1985–86.

It is widely regarded as the tightest hole ever drilled. The rumor mill has ground incessantly over the well, which is thought to have encountered hydrocarbons. Some unsubstantiated reports told of daily flow rates in the tens of thousands, perhaps multiple tens of thousands of barrels. They should be taken with more than a grain of salt. Neither Chevron nor its partner in the well, BP, has ever given an inkling as to whether the well ever encountered anything. That would be a pointlessly stupid and self-destructive act from a competitive standpoint. Should ANWR ever be leased, Chevron and BP will give away a proprietary advantage to their competitors in what could be one of the biggest finds ever in North America. Even if the well proved dry, the two companies still would have an edge in valuable geological knowledge about the Coastal Plain that would be critical in leasing elsewhere in the area. Yet, the fruit of their efforts probably won't be known until there is leasing on the Coastal Plain. No further drilling is allowed in ANWR, even for exploratory purposes, and even on native lands, development remains forbidden. As it was with the development of Prudhoe and the building of TAPS, it will take an act of Congress to find out what really lies within the most prospective untapped petroleum province in North America.

ANWR Issues

The debate over access to ANWR encompasses a variety of environmental issues. At the core is the debate over the concept of wilderness value itself.

In late summer 1989, environmentalist groups sought to compete with a media tour of ANWR and the Exxon Valdez cleanup sponsored by the American Petroleum Institute and Alaska Oil and Gas Association by staging a selective media tour of their own. While helicopters ferried some reporters to the site of the 1 KIC Jago River well and distinctive seep and outcrop sites in the 1002 area, the environmentalists chose a select few reporters to take a wilderness expedition along a major river coursing through the heart of the wildlife refuge.

Media on the environmentalists' tour got a rare glimpse of a spectacular, pristine wilderness teeming with wildlife. Media on the industry tour mostly saw a flat, treeless plain dotted with disconnected small ponds, a few seabirds near the seashore, and two or three caribou and visited a small native village.

The distinction pointed to the environmentalists' strategy of attempting to disguise the fact that there are actually two ANWRs. One is a spectacular wilderness that comprises 92% of ANWR south of the Coastal Plain and contains mountainous regions extending to the southern edge of the Brooks Range.

More than half of ANWR, including a 50-mi stretch of the Coastal Plain, has already been designated wilderness, and is thus forever off-limits to resource development of any kind. The other ANWR is the one that the petroleum industry is interested in: the 1.5 million-ac Coastal Plain that contains North America's most prospective untapped petroleum potential.

Environmentalists contend that leasing and drilling in the ANWR Coastal Plain would not just destroy the untouched wilderness value of the wildlife refuge, but it would also harm the ecosystem there and damage the habitat for wildlife, irreversibly. Any development would cause dramatic drops in the number and distribution of many species that use the area to reproduce, they contend. They also maintain that the Coastal Plain itself is a unique example of arctic wetlands not found anywhere else in America.

However, the wilderness argument regarding ANWR overlooks many salient facts, not the least of which is the contribution Alaska has already made to the U.S. wilderness resource. Under ANILCA alone, more than 104 million ac of Alaskan lands were withdrawn into federal conservation units. Altogether, more than 158 million ac in Alaska have been set aside as federal conservation units. In sum, these lands, about equal to the combined size of California and Oregon, account for 70% of America's national park lands and 90% of its wildlife refuges.

Of that huge amount of set-aside lands in Alaska, the federal government already has designated 57 million ac as wilderness areas. Under that designation, all types of resource development, including commercial fishing, is prohibited. A wilderness designation also outlaws construction of public and private recreational facilities such as small backcountry lodges, ski resorts, visitor centers, and motor tours. Except in a few rare instances, roads, new access trails, and cabins are forbidden in federal wilderness areas.

In other words, a federal designation of wilderness would deprive the vast majority of Americans access for even the simplest of recreational activities. The environmentalists speak of wilderness as a resource that is to be part of America's heritage, enjoyed and used by future generations. But in truth, a federal wilderness designation would render an area generally off-limits to most families, the elderly and handicapped, and persons of limited financial means.

It follows that, for the most part, the only human users of federal wilderness resources will be those affluent, younger professional types who are the core of the membership of the Sierra Club and other groups and are often heard to complain about "the Winnebago set" crowding "their" vistas. There is in this an ugly undercurrent of elitism that carries an even uglier subtext of ageism and racism that cannot be easily dismissed by invoking the usual allegiance to liberal orthodoxy. You just are not going to find many blue collar families, lower-income minority groups, or 80-year-old stroke victims kayaking along the Yukon River or backpacking in the Brooks Range.

When the environmentalist extremists decry the presence of man as damaging to the wilderness aesthetic, invariably it is accompanied by an unspoken, "Except for me." Would it not be an extension of their logic regarding wilderness to exclude the presence of *all* humans from a wilderness, including the natives who subsist there? Even a subsistence lifestyle in Alaska is hardly primitive; in Alaska, the fishing and hunting generally involve outboard motors, snowmobiles, and the latest weaponry and fishing gear.

And what of the backpackers and kayakers and climbers? Is their presence not an intrusion as well? I've backpacked alone in Alaska, and all the sourdoughs advised this "cheechako" to make as much noise as possible hiking in bear country or risk an abrupt and unpleasant encounter with a startled grizzly sow, whose protective instincts toward her cub extend the breadth of a grizzly's range (as much as 100 mi) and who does not give one whit whether your Sierra Club dues are fully paid up. Does it not follow, then, that any presence of human beings constitutes a violation of the pristine, untrammeled nature of wilderness? Shouldn't the Sierra Clubbers and Audubon Society folks forego forever their "back to nature" experiences—not to mention the hordes of photographers that descend upon these "untouched" regions of the earth every year to provide grist for their magazines, calendars, coffee table books, and fundraising brochures (often printed on non-recycled paper and with inks derived from petroleum)—as unwarranted intrusions upon the wilderness aesthetic?

The philosophy behind the wilderness aesthetic as defined by the environmental lobby aside, the opponents of environmental leasing have claimed that three decades of North Slope exploration, drilling, and production have left behind a legacy of environmental devastation. They cite as purported evidence of this a draft report by staff at Interior in early 1989 that alleged such damage to the North slope ecosystem by industry operations. Among the claims was the supposition that an oil spill may have killed about 60 acres of tundra. Such reports formed the basis for environmentalists' claims that any development on the neighboring ANWR Coastal Plain would cause dramatic drops in the number and distribution of species that use the area to reproduce.

However, biological studies covering more than three decades of exploring, drilling, and producing oil on the North Slope have yet to yield any evidence that oilfield activities have produced any measurable change in the population size of any fish or wildlife species using the North Slope. The best rebuttal to such claims lies in the fish and wildlife populations that continue to feed, reproduce, and raise their young throughout the development areas of the North Slope. And there is not a shred of evidence to show that the ability of habitat to support fish or wildlife species in the area has been affected by the fraction of tundra—less than 2%—occupied by North Slope oil activities.

Many of these reported problems at Prudhoe Bay have resulted from poor operating practices by oilfield service contractors based at the town of Deadhorse, according to North Slope field operators. Some of those contractors have abandoned wastes at their sites, which are on lands leased from the state. North Slope operators have stepped in voluntarily to clean up these problems. At the adjoining Kuparuk development, service contractors have been consolidated in an industrial center and placed under more direct control. That will be the case at ANWR as well.

Although the industry has repudiated these claims, EPA has noted that the claims have been made and subsequently addressed before, and Interior itself has discredited the draft report; the opponents of ANWR leasing nevertheless continue to cite the draft report in their hysterical fundraising brochures.

The environmentalists claim that leasing and development would destroy the wilderness character of the Coastal Plain. However, the stretch of Coastal Plain east of the 1002 area contains 450,000 ac of federally designated wilderness. This region includes a complete continuum of terrain and types of habitat like those within the 1002 area, as well as five major river systems.

That argument also lies behind the rationale of legislation that would call for a limited approach to ANWR Coastal Plain exploration. But such an approach would impede and perhaps prevent discovery of a Prudhoe Bay equivalent. The bill specified that only four wells be drilled on the Coastal Plain in order to assess its "true" potential—presumably with the notion that if these four wells did not yield a commercial success, the United States could somehow feel justified in writing off ANWR's Coastal Plain as a petroleum province. But the Coastal Plain has a geologic complexity unlike that at Prudhoe, where it nevertheless required 10 wildcats before a discovery was made, not to mention the dozens of wildcats drilled elsewhere on the slope in the years before Prudhoe. The Coastal Plain has a geologic complexity unmatched on the North Slope. At least 26 distinct, separate geologic structures have been identified on the Coastal Plain. It could take 40 wells to turn up a commercial strike, and that strike might well be another Prudhoe Bay.

Another red herring that the environmental lobby has used in opposing ANWR oil work is the notion that leasing and development would eliminate large areas of habitat from the 1002 area. In fact, even with development of three large oil fields in the area, less than 1% of the 1002 area's land surface would be affected by direct habitat removal and by ancillary effects such as vehicle road dust.

Environmentalists wield the term "zone of influence" to describe how oil and gas development might affect fish and wildlife populations through direct habitat loss, displacement of wildlife from habitat, or blocking wildlife from their habitat. The first example has never been shown, and, in fact, is contradicted by the persistent growth in wildlife populations in existing North Slope development areas. The latter two processes have never been documented on the slope and therefore remain hypothetical.

In any event, the next phase of North Slope development, should it occur in ANWR, will result in a significantly smaller "footprint" on the tundra by industry activities. Advances in the petroleum industry's drilling and production technologies have enabled North Slope operators to make great strides in reducing their industrial presence in the sensitive arctic region.

Arctic wells and facilities are placed on gravel pads that insulate the permafrost—a permanently frozen subsoil underlying the tundra as deep as 2,000 ft—from thawing. With advances in directional drilling technology, the size of new wellpads has been cut sharply. With many wells drilled directionally at high angles from a single pad, hundreds of wells can be drilled from just a handful of pads. A typical drillpad used in early development at Prudhoe Bay field covered about 49 ac and was designed for 32 production wells. The well spacing then was usually 120 ft or more. By the time development began at Kuparuk River field, spacing was as close as 60 ft. Industry has advanced to the point where well spacing can be as little as 10 ft, the case at Endicott offshore field just 2 mi from the Prudhoe facilities.

Further, drillpads on the slope used to include reserve pits big enough to contain drilling muds, cuttings, and fluids from all the wells drilled on that pad. Now North Slope operators design reserve pits to contain only the solid cuttings from a well, reinjecting the produced fluids or injecting them into a waste disposal well.

As a result of these advances, a drillpad envisioned for ANWR would contain twice as many wells and cover less than half the area as that at Prudhoe.

The same situation holds true with other North Slope facilities. Production facilities at ANWR could be built using half the surface area as that used at Prudhoe. The consolidated service center at ANWR, like that at Kuparuk, would similarily occupy a fraction of the service center at Deadhorse: 55 ac compared with 1,050 ac.

Consequently, the total surface area occupied by development of a large field and a small field on ANWR's Coastal Plain would total only about 5,000–7,000 ac—less than 1% of the 1002 area.

And, industry no longer builds gravel roads to construct pipelines, as it did early in Prudhoe development. The approach that would be used in ANWR would be the same as that in Endicott development. There, the pipeline was constructed entirely from an ice road through the Sagavanirktok River delta. When the weather warmed, the ice road melted, leaving no trace and thus no disturbance to the ecosystem.

Opponents of ANWR leasing also claim that development would further worsen North Slope air quality already degraded, they say, by years of industry operations at Prudhoe Bay. However, the average annual concentration of sulfur dioxide on the North Slope is below detectable limits. The only air pollutant of concern, then, is nitrogen oxides. Environmentalists cite air pollution levels on the North slope as bad as in some urban areas in the Lower 48. Those claims apparently rested on the level of maximum allowable emissions covered under government permits, which for nitrogen oxides at Prudhoe Bay are as much as 100,000 tons. But the actual emissions of nitrogen oxides at Prudhoe average less than 20,000 tons.

Environmentalists also charged that Prudhoe operations have contributed to arctic haze and acid precipitation. The National Oceanic and Atmospheric Administration in 1986 found that Prudhoe oil operations did not contribute to arctic haze, which since has been traced to industrial pollution carried over the arctic from the middle latitude regions of Europe and Asia. The acid rain charge is equally spurious, with an undetectable level of sulfur dioxide and an unmeasurable contribution from nitrogen oxide.

ANWR leasing supporters also have had to refute claims that drilling and production waste management practices at Prudhoe have fouled the environment. The main argument is that production wastes placed in permitted reserve pits built onto the gravel drillpads leak fluids that have caused widespread environmental damage. Although there have been some pit wall failures and discharge onto the tundra, there remains no evidence of significant environmental impact. And EPA has determined that drilling fluids and cuttings are exempt from regulation as hazardous wastes.

Perhaps the biggest environmental concern involving access to ANWR's Coastal Plain is that involving the caribou. Although industry thought it had laid the issue to rest with the completion and safe operation of TAPS, environmentalists resurrected fears over industrial development causing destruction of habitat, disrupting caribou calving grounds, or blocking caribou migration.

They contend, in essence, that petroleum operations in ANWR would displace caribou from large areas. The same argument was in opposition to Prudhoe Bay development and construction of TAPS. In a classic case of selective memory, environmentalists testifying at federal hearings on TAPS even suggested, in opposing the TAPS route, that the pipeline instead be built through what is now ANWR, contending the latter had little significant wilderness or wildlife resource value. Countless studies at the time and since have shown that the caribou remain generally unaffected by North Slope oil work and TAPS. In fact, the caribou often huddle beneath the heated pipeline for warmth and seek out gravel pads as havens of insect relief. The best argument that can be made regarding the impact of oil installations on calving grounds is that the caribou persist in calving in the areas where, it is claimed, disturbances would bar them—Kuparuk River and Milne Point fields.

The caribou herd in the Prudhoe Bay area has grown from about 3,000 in 1972 to more than 18,000 today. In that time—spanning North Slope development and the building of TAPS—there has yet to be ascertained a detectable effect on regional distribution, calving success, patterns of migration, herd size, productivity, or anything else of note.

The zeal to protect caribou movement from hindrance led TAPS builders to be saddled with having to build caribou "ramps" over pipeline sections at a cost of about $1 million apiece to assuage concerns that caribou would not walk under the raised pipeline and therefore would be cut off from part of their habitat. It is an article of faith among North Slope veterans that many, many more caribou walk under, graze beneath, and nuzzle the pipeline than those who use the ramps.

174

This mountain of evidence supporting industry's sparkling environmental record on Alaska's North Slope notwithstanding, environmentalists insist that ANWR's postulated oil and gas resource is not enough to justify the environmental risk. They further argue that alternatives such as conservation or improved fuel efficiency standards would offset the need for ANWR oil.

For example, a widely cited figure among environmentalists is that the estimated mean potential of 3.2 billion bbl underlying the ANWR Coastal Plain would satisfy only 2% of U.S. demand during a hypothetical 30-year life of the field. This is a specious argument. What the environmentalists choose to overlook is the one in 20 chance of there being between 26 and 29 billion bbl of oil on the Coastal Plain, about the same as the oil in place at Prudhoe Bay. Were it not for Prudhoe—and the development of fields like Kuparuk River, Lisburne, Endicott, and Milne Point, fields whose economics were made viable only by the Prudhoe infrastructure and TAPS—the U.S. level of dependence on foreign oil imports would have exceeded 50% long before now— perhaps even during the price spikes of 1979–80.

If the upper estimate of Coastal Plain potential proves correct, and projections of U.S. production declines are accurate, it is not too big a speculative jump that ANWR's Coastal Plain and remaining North Slope production might provide more than half the nation's domestic supply of oil well past the turn of the century (Fig. 7–2).

There is a critical argument for developing ANWR oil on behalf of a key minority group. The Inupiat Eskimos, who initially opposed development at Prudhoe Bay, have come to have their lives transformed for the better by environmentally sound oil development on the North Slope. Money from oil development has enabled the Inupiat to upgrade their lives significantly from what had been an arctic ghetto to one marked by good schools, medical care, electricity, running water, and sanitation. Still, the Inupiat depend on a basic subsistence lifestyle, hunting and fishing to eat, while maintaining their cultural traditions. Allowing ANWR Coastal Plain development would sustain the cash economy that allows the Inupiat to continue their subsistence lifestyles. Designating the Coastal Plain as a wilderness would condemn the Inupiat to living on welfare, not to mention the wilderness designation complicating subsistence activity.

It is no secret that the subsistence activities of the Inupiat, who are the only Americans allowed to harvest whales, have long stuck in the craw of the most strident environmentalists. Putting the Inupiat on the dole, extending

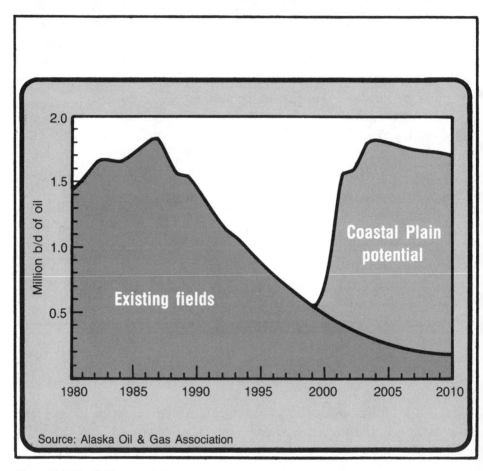

Figure 7–2 North Slope outlook.

the American welfare state to the frigid north, would conveniently put a halt once and for all to the hunting of whales in American waters. Again, the paternalistic sanctimony of the environmentalists has a slight tinge of racism to it.

There are arguments for ANWR oil contributing to the U.S. economy as well as those for energy security. The biggest single component—and the most volatile one—of the U.S. trade deficit is imported oil, by far. Interior estimates that a discovery of even 3.2 billion bbl of oil could result in more than 250,000 new jobs and $79 billion in economic benefits. If ANWR holds as much as 9.2 billion bbl, Interior estimates, the net economic benefit may total as much as $261 billion. ANWR oil will contribute income from lease bonuses, rent, and royalties, to the federal treasury—one of the few sizable, politically acceptable revenue sources to help deflate the ballooning federal budget deficit.

In 1988, the U.S. trade deficit totaled more than $171 billion. Of that amount, almost $45 billion was spent on imported oil. If current trends continue, the United States could be importing $150 billion of oil each year by 1995—$15 billion more than the entire trade deficit for 1986—and that is at conservative price forecast scenarios, which are dwindling in number and surety these days. In 1987, Alaskan oil production trimmed the U.S. price tag for imported oil by more than $12 billion—in a year when oil prices often hovered near historic lows, adjusted for inflation.

Those economic benefits from Alaskan arctic oil production are not limited to Alaska or the federal government. During 1980–86 alone—after the completion of TAPS and start-up of Prudhoe Bay field—money spent on North Slope development in the United States totaled more than $10.5 billion, excluding about $7 billion spent on intangible items for Kuparuk, Prudhoe Bay, and Lisburne oil fields. Every state shared in that bonanza, from $3.5 billion spent in Texas and $1.8 billion spent in California to $300,000 spent in New Hampshire and $600,000 spent in Rhode Island.

Since 1974, oil companies have spent more than $45 billion on North Slope development and TAPS construction. Certainly they have reaped billions of dollars in revenues from this massive investment; but the economic benefits have filtered down through every sector of the American economy, not just to corporate shareholders. And as America grapples with increasingly intractable deficits and the other economies of the world become more and more competitive, that ultimately might prove a more compelling argument for access to ANWR Coastal Plain oil than the sometimes fuzzy geopolitical scenarios of energy security.

There are other environmental issues that not only could affect ANWR Coastal Plain exploration and development, but also pose a threat to exploration and development elsewhere in the state. Alaskan explorationists may have dodged a big bullet in the wetlands controversy early in 1990. The environmental lobby has been pushing to stem the loss of America's wetlands by urging Congress and the president to implement a policy of "no net loss" of U.S. wetlands. Half of the nation's wetlands have been lost to agriculture, industry, and real estate. This policy would halt that decline by seeking to block any economic activity compromising existing wetlands unless the promoter of that activity compensates that loss by creating or preserving a comparable wetlands resource elsewhere.

Opponents of ANWR Coastal Plain exploration and development have sought to exploit this concern for their cause. However, the situation in Alaska differs greatly from the Lower 48's declining wetlands resource. Alaska holds about 170 million ac, which comprise more than 60% of the nation's total. There remain 37 million ac of wetlands on the North Slope—bigger than the states of California and Oregon combined—of which only 20,000 ac have been affected by oil activity. In all, 99.95% of Alaska's wetlands remain untouched.

Further, North Slope wetlands are different from those in Alaska below the Arctic Circle. Subarctic Alaskan wetlands hold as many as 100–250 waterfowl/ sq km. On the North Slope, that density drops to less than 5–10 waterfowl/sq km. Even at that, waterfowl concentrations within the North Slope oil fields have remained stable or increased since before development activity began.

There also is a fundamental difference in the character of arctic wetlands versus those in the Lower 48. North Slope wetlands are wetlands by default, not by climate or related water systems, as elsewhere. Technically, the area is regarded as an arctic desert, with precipitation averaging less than 10 in./yr. Because the top layer of the permafrost underlying tundra melts about 1 ft during the eight weeks or so of arctic summer, it causes snowmelt and rain to be trapped on the surface, thus resembling true wetlands in the Lower 48.

In other states, wetlands contribute to wildlife habitat, charging aquifers, controlling erosion, containing floodwaters, protecting shorelines, and providing outdoor recreation. That is generally not the case on the North Slope, and what little wildlife habitat there is is provided only briefly for migratory birds.

For now, the issue may be moot. President Bush has backed off his "no net loss" pledge with an agreement in March 1990 between EPA and the Army Corps of Engineers that would allow certain exceptions to the rule. The issue certainly will arise again, as general policy, and with application to specific projects in Alaska. If the "no net loss" rule becomes strictly interpreted and widely implemented, it probably will kill any new exploration and development—not just on the North Slope, but also on many other promising areas of Alaska. One approach that would be advisable for industry is to pursue the concept of wetlands mitigation banking—creating, restoring, or preserving wildlife habitat when a proposed development calls for loss of wetlands. This is the future direction of sound wetlands policy.

Another sticky problem for North Slope oil explorationists and producers developed in the Army Corps' initial opposition to installation of more gravel causeways in nearshore waters off the North Slope coast. Such causeways were essential in the development of Prudhoe and the other North Slope fields, and a ban on them could kill new oil and gas field development on the North Slope, thus rendering exploration efforts pointless.

The controversy has existed since the early days of Prudhoe development, when operators installed the first causeway for the West End Dock at Prudhoe Bay. As would be the case on ANWR's Coastal Plain, the dock at Prudhoe was critical for receiving giant barges laden with supplies, equipment, and the monster modules destined for Prudhoe Bay field. Nearshore Beaufort waters are so shallow that barges must offload several miles out. Concerns were raised that the gravel causeways interfered with fish migration patterns. The corps remains concerned about changes in salinity and temperature in nearshore waters caused by the causeways. The North Slope operators sought to alleviate those concerns by installing huge breaches in the causeways for the Prudhoe seawater treating plant and Endicott development project. The breaches, although costly, allowed fish to pass through as if the causeways were not there.

The issue was resurrected when BP Alaska sought a permit for its proposed development of the marginal Niakuk field near Prudhoe Bay, a "small" (by Alaskan North Slope standards, at 51 million bbl of oil reserves) offshore accumulation. The corps denied BP a permit partly because of a procedural error in denying the permit, and partly because pressure by the departments of Interior and Energy noted that the permit denial blocked development of 60% of the oil discovered in the United States in 1987.

The corps also threatened to shut down Endicott field with warnings that the project's causeway was causing harm to the environment and required a $40-million retrofit. Since then, the corps has backed off its original tough stance and has been accommodating some sort of compromise—with further studies and monitoring underwritten by industry—on the causeway issue.

In short, without causeways, there probably can be no further North Slope development. Without the opportunity to develop, there is no reason to explore. And without access to ANWR's Coastal Plain and other crucial areas of the most prospective region in the United States, there is no sense in even pretending a serious effort at promoting a coherent National Energy Strategy exists.

Offshore's Promise

• •

After ANWR's Coastal Plain, the OCS offers the best—perhaps the only really significant—promise of discovering giant oil fields in the United States.

The OCS contains more than half of the oil remaining to be discovered in the United States, according to the United States Geological Survey. Outside ANWR's Coastal Plain, there is no onshore U.S. basin with even remotely the kind of untapped hydrocarbon potential that some of the offshore provinces contain.

Estimates of U.S. offshore oil and gas resources vary widely. A useful estimate might be the roughly 29 billion bbl of oil remaining to be discovered in federal waters off U.S. coasts, according to the latest USGS data. If typical industry recovery rates are applied, that works out to another Prudhoe Bay equivalent in terms of reserves. Clearly, the majority of those undiscovered resources lie in the areas that are the subject of environmental concerns. That is because, outside the central and western Gulf of Mexico, *all* U.S. OCS areas are the subject of intense environmental concerns. In the heavily drilled shallow waters of the central and western Gulf of Mexico, the remaining prospects of interest tend to be gas prone. The oil prospects of interest in the Gulf of Mexico are primarily in the deep and ultradeep waters, and thus carry with them technological and economic constraints where the environmental constraints might be manageable.

Where there are prospects for giant oil fields to be found offshore, they tend, unfortunately, to be off the most acutely ecologically sensitive areas: offshore Alaska, California, and Florida. These are the areas in which the petroleum industry has evinced the greatest interest. For example, off California, the Interior Department estimates recoverable oil resources could total 2–5 billion bbl. However, some industry estimates put that potential at perhaps as much as 10 billion bbl of recoverable oil. That estimate is just for the federal lands on the OCS. State tidelands within 3 mi of shore—and thus not under Interior's jurisdiction—could hold several billion bbl more, given the generally better geological conditions near shore.

Under the Outer Continental Shelf Lands Act Amendments (OCSLAA), Congress issued a mandate to Interior: that the agency must manage the resources on the OCS for the maximum benefit of Americans, notably through leasing those lands for exploration and development of mineral resources, chiefly oil and gas.

To comply with this mandate, Congress instructed Interior to devise a five-year schedule of oil and gas leasing on the OCS, taking into consideration local input from the areas affected, chiefly with regard to environmental concerns. Interior has dutifully attempted to comply with that mandate through the years, with decidedly mixed success. Although leasing accelerated sharply during the Reagan years, especially with the advancement of the areawide leasing concept, the political opposition to leasing certain key areas, such as Offshore California, has kept pace.

The rapid pace of OCS lease offerings during the Reagan years has slowed substantially. It was largely in response to the atmosphere of contentiousness surrounding Reagan's first Interior secretary, James Watt, that Congress responded with a series of lease sale moratoriums it built into the agency's annual budget appropriations process. That has persisted now for nine years—fiscal years 1982 through 1990. The areas covered by the congressional moratoriums have varied from year to year. But the total has jumped from only 736,000 ac to more than 84 million ac, because the moratoriums cover the most hotly disputed areas—off California, Alaska's Bristol Bay (North Aleutian) basin, off Florida's gulf and east coasts, and off most of the Atlantic coast.

The tenor of congressional moratoriums has changed as well, especially since the Exxon Valdez oil spill. Previously, those moratoriums had tended to be defensive in intent, a response to an aggressive lease-sale schedule. Now, they tend to take the offensive. Typical of that approach is a bill filed in early 1990 by Rep. Barbara Boxer (D.-Calif.) and 24 other congressmen. The bill would establish drill-free zones around certain coastal states, effectively making existing moratoriums permanent and creating new ones as well. It would ban drilling 50 mi off Massachusetts; 100 mi off Florida, Georgia, South Carolina, Delaware, New Hampshire, Maine, Oregon, Washington, and Alaska; 125 mi off New Jersey, New York, Connecticut, and Rhode Island; 145 mi off California; and 175 mi off North Carolina. Conspicuous by their omission from the drill-free zones are Texas, Louisiana, Mississippi, and Alabama. Apparently, these states are not sufficiently deserving of the level of environmental "protection" that would be afforded the other coastal states. The bill also would require the government to buy back leases on acreage off Florida and in Alaska's Bristol Bay.

This stepped-up opposition to OCS leasing has taken shape since the Exxon

Valdez spill. It follows, then, that Minerals Management Service—the agency within Interior responsible for all oil and gas leasing—plans to dramatically change its approach to leasing. MMS will schedule fewer sales, offer less acreage, and pay more attention to environmental concerns. Explorationists can expect lease sales to have environmental stipulations tailored specifically for those areas. And, there will be more emphasis on early decision-making in the sale process itself as MMS works more closely with local coastal communities.

Areawide leasing, while generally praised by the industry during the 1980s, will pretty much be a thing of the past outside the Gulf of Mexico. A much more selective approach will be used for acreage offerings in the future. Although that won't necessarily reduce the amount of acreage actually leased that drastically—since only a fraction of the vast amount of acreage offered under the areawide concept was leased—it is probably safe to assume that a selective approach will certainly limit companies' opportunities to bid because some of the geological play concepts will be eliminated in the process.

Those changes will be reflected in the next five-year lease sale schedule, which was delayed in spring 1990 in order to wait for President Bush's decision on two sales planned off California and another planned off Florida and for the DOE's draft report on a National Energy Strategy. In early 1989, Bush named a White House task force to assess environmental concerns related to the three controversial sales. In the process, he put those three sales—originally scheduled for 1990—on hold. The task force responded without a specific recommendation. Instead, it gave Bush a series of findings and options. That development alone should have alerted explorationists to the certainty of a debate over OCS leasing that would be marked by hyperbole and emotion and not science and fact.

The task force report generally followed the line of a study it requested from the NAS in 1989. NAS concluded in its study that Interior did not have enough environmental and socioeconomic data to justify oil and gas leasing and development in the three sale areas.

In other words, the government does not have sufficient scientific and technical information available about the potential environmental effects of offshore oil and gas leasing in order to make sound decisions on drilling and production. The NAS report offered no opinion as to whether the sales should be held after more data is gathered.

That, in turn, has called into question the entire process of how Interior has evaluated the environmental impact of offshore exploration, development, and production during the past 20 years—and the quality of the data that process has produced. That could have an especially chilling effect on future lease sales because it offers a strong point for a legal challenge. Anti-leasing forces ultimately have not fared well in the courts when it came to efforts to halt lease sales. They have succeeded, however, in delaying sales through litigation. Even the longest delay, however—legal action that blocked MMS from opening Bristol Bay leases for almost two years—ultimately was unsuccessful.

Another concern raised by the NAS report was that there appeared to be no separation in MMS decisions on whether to lease tracts and whether to permit development after a lease has been awarded. In other words, because Interior has never denied a development/production plan—although some have been changed—that has led the public to conclude that leasing automatically means development. That dichotomy hinders the processes of effective environmental assessments and credible public dialogue, NAS contended. It recommended as an alternative a more comprehensive environmental impact statement at the development and production stage and called on MMS to make an effort to split decisions regarding leasing/exploration and development/production with distinctly different phases of data gathering and analysis.

That may sound reasonable at first, but what company would be willing to spend the money, time, and effort to pay for a lease (some in the tens of millions of dollars—and in the boom years of 1979–81, hundreds of millions of dollars) without some assurance that it be allowed to develop in the event of a discovery?

Although there is ample opportunity for data gathering and public input throughout the EIS process, explorationists can expect MMS to develop a more distinctly phased approach in its OCS program during 1991–96. One strong likelihood is that Interior will have MMS extend the lease sale schedule to 10 years from the current five in order to allow more time for environmental and other prelease studies. Furthermore, different areas of the OCS targeted for leasing under the next lease sale schedule are likely to have different treatment in the EIS process. No doubt the "worthier" coastlines of California and Florida will get a much more protracted EIS process than those of Texas and Louisiana.

In any event, extending the schedule to 10 years violates the congressional mandate under OCSLAA. And, giving different areas different treatment flouts the spirit of OCSLAA, which calls for an equal sharing of the environmental impacts among coastal regions (in practice, that has never been the case, anyway). The intent has always been to maintain a geographical balance. It is logical that that geographical balance will shift as some OCS areas do not live up to their advance billing after initial exploration. Some sales are dropped for lack of industry interest following unsuccessful exploratory campaigns.

However, some very prospective areas, most notably California, have managed to avoid leasing mainly because of the environmental lobby's political clout there. It is no coincidence that, of the eight lease sales scheduled originally in 1989 under the current five-year leasing plan, only two were held—in the central and western Gulf of Mexico. Bear in mind that 1989 sales scheduled off northern California and the northern Atlantic states were canceled or delayed in February, *before the Exxon Valdez spill.* The last five lease sales held were in the Gulf of Mexico, and the only lease sale held in 1990 was one in the western gulf in August.

In light of the Exxon Valdez, then, perhaps it should not have come as a surprise that President Bush in late June 1990 went beyond the NAS recommendations and canceled eight disputed OCS sales, preventing drilling in those areas until after 1990.

His action essentially killed all pending lease sales off California, calling for more studies in the next three to four years. An exception would be 87 high resource potential tracts in the Sale 95 area in the Santa Barbara Channel and Santa Maria basin that may be offered after 1995 if studies are favorable. Even for this paltry (less than 1% of California's offshore area) area, don't expect leasing without a fight. Bush also approved a proposal to create 2,200 sq mi marine sanctuary in the Monterey bay area—permanently off limits to leasing.

Bush also canceled until after 2000 a Gulf of Mexico sale southwest of southern Florida, pending completion of more studies. He further ordered MMS to cancel and repurchase existing leases there.

In addition, the president canceled two North Atlantic sales, ordering MMS to not schedule a sale there in the next five-year leasing plan or before 2000. He also canceled sales off Washington and Oregon, calling for another five to seven years of studies.

Bush's actions clear the way for MMS to proceed with a new five-year leasing plan, but leave very little outside the Gulf of Mexico and selected noncontroversial Alaskan basins on the current five-year schedule (Table 7–1).

Tentatively, at least, MMS will issue its draft lease sale schedule in November 1990. After comments are reviewed, the agency's final program will be issued

Table 7–1 Remnants of the Five-Year OCS Lease Sale Schedule*

Date	Sale Number	Area
1990		
August	125	Western Gulf of Mexico
September	SU2	Supplemental
1991		
January	120**	Norton basin
February	124	Beaufort Sea
March	131	Central Gulf of Mexico
May	126	Chukchi Sea
July	121**	Mid-Atlantic
August	135	Western Gulf of Mexico
September	130**	Navarin basin
	SU3	Supplemental
November	137	Eastern Gulf of Mexico
1992		
Early	108**	South Atlantic
March	139	Central Gulf of Mexico
May	133**	Hope basin
On hold		
	107**	Navarin basin
	101**	St. George basin
	117	North Aleutian basin
	114**	Gulf of Alaska/Cook Inlet

* Following President Bush's cancellation of eight sales in midsummer 1990. ** Frontier sale, to be held only if industry shows enough interest.

in November 1991. Congress then has two months in which to accept or reject the program.

Absolutely critical to the fate of OCS leasing in the 1990s will be DOE's National Energy Strategy. Both opponents and supporters of OCS leasing will seek in that policy documentation vindication for their respective claims. Environmentalists already have put forth the argument that an NES is required because it ostensibly will illustrate that offshore drilling will accomplish little in enhancing the nation's energy security and thus is outweighed by environmental risks. Explorationists, of course, believe an NES is necessary to prove the opposite. What it is likely to contain—as have virtually all of the Bush administration's energy/environment policy dispute resolutions—is a sort of middle path that touts the benefits of conservation and alternate energy sources, pushing for offshore drilling where the political heat does not overwhelm the potential resource, and tightening environmental strictures across the board in energy resource exploration, development, and production.

In other words, expect more lease sales and drilling in the Gulf of Mexico, and selected basins off Alaska like the Beaufort Sea, Chukchi Sea, and Cook Inlet (and perhaps the St. George, Norton, and Navarin basins in the Bering Sea, if industry's interest has not flagged there). Don't count on a lease sale off California or any of the other areas targeted by Bush in the 1990s unless there is the very real prospect of sharp and substained price spikes owing to some oil supply shortfall.

The irony of all of this is that blocking leasing and development of offshore oil and gas resources actually creates a much greater threat to the environment than allowing it. That is because the U.S. economy requires a low-cost energy resource to remain competitive, so it will simply get the oil it needs via foreign flag tanker. As tanker traffic to U.S. ports increases, that will sharply boost the risk of a catastrophic oil spill. In the past half-century, almost all major oil spills resulted from tanker accidents. Only a small percentage of U.S. offshore oil production moves to market via tanker.

The U.S. offshore industry has produced almost 5 billion bbl of oil since 1975 and lost a mere 900 bbl due to a blowout. That works out to 0.0000002% spilled from drilling-production operations. To this day, there has never been a single barrel of oil spilled off the United States because an exploratory well blew out. And environmental and safety rules have gotten only stricter in that time.

Since 1970, only 160 bbl of oil have spilled as a result of accidents involving offshore platforms off California. In contrast, natural oil seeps along California's coast discharge as much as 20,000 bbl/yr into those scenic waters. That does not include the huge volumes of much more toxic oil products routinely discharged off California with the blessing of the state's agencies. Since 1975, municipal wastewater treatment plants between Santa Barbara and Newport Beach have discharged more than 4 million bbl of oil and grease into waters off California, not to mention millions of tons of other pollutants. By comparison, oil discharges from petroleum industry operations don't even register statistically. Worldwide, of all the oil detected in the oceans by a 1985 study of the National Academy of Sciences, only 2% came from offshore oil operations.

But science and logic are taking a back seat in this controversy. That can be seen in a failed amendment by Senator McClure in summer 1989 to the

Interior appropriations bill (which imposed the moratoriums) that would also ban oil tankers from traversing the moratorium areas. McClure contended that if the goal of the moratoriums is to protect the environment, then the proper response to the rash of tanker spills that prompted the moratoriums is to ban those tankers from the sensitive areas.

"The most important fact, however, is that none of those disasters had anything to do with drilling in federal waters. The Outer Continental Shelf drilling program does not have a safety problem," McClure said.

"If the goal of placing a moratorium on OCS development is to protect the environment, then let's stop fooling the American people. Tankers are more dangerous than oil rigs, so let us eliminate tanker traffic."

Beyond efforts to stymie the OCS lease sale program, states and local municipalities have waged their own wars to block offshore oil leasing, directly and indirectly. The efforts to hinder development of giant oil fields off California by environmental groups, public agencies, and politicians have been so successful that companies like Chevron are backing off plans for development of other fields in the future. The anti-oil atmosphere has gotten so heated there that, even if through some miracle a lease sale were held off California soon, it's possible that no one would come.

Although Offshore California is arguably the best offshore prospect in the United States (prospective structures off Alaska may be larger, but the likelihood of a discovery and commerciality for a field is greater off California), few companies would be willing to suffer what Exxon, Chevron, and ARCO have in attempting to win approval for their development projects off Santa Barbara County. Their sagas will be covered in the next chapter. Even Interior Sec. Manual Lujan, whose mandate it is to maximize the use of OCS resources, in 1989 predicted he will be "long gone from Interior before we lease one more acre of land off California."

Although federal courts have consistently upheld the legality of Interior's five-year lease sale schedule (most recently a couple months before the Exxon Valdez spill), state and local governments can find other means to hamstring exploration off the nation's coasts. For example, a number of cities and counties along California's coast have adopted municipal bans on the installation of onshore facilities related to offshore drilling activity—including even the manufacture, sale, or provision of equipment or supplies for OCS oil work. That creates a real barrier for development, thus rendering a lease almost useless if a legal challenge is unsuccessful. And states have been equally successful at delaying lease sales through court challenges of environmental impact statements' adequacy for individual lease sales.

Furthermore, there is the mandate of the Coastal Zone Management Act, which requires federal agencies to address concerns of coastal states and communities in all permitting related to offshore work—deeming that work be "consistent" with local concerns. Efforts to use the CZMA consistency powers vested with state agencies to block OCS leasing failed. However, agencies such as the California Coastal Commission have been so ruthlessly efficient in using their permitting powers to block OCS exploration and development that the initial right of access to some areas may simply become moot.

In effect, that gives a state, county, or city veto power over use of a national resource. And, since the Exxon Valdez, that veto power will be used with greater frequency. Industry must undertake the most effective program of commitment to environmental safeguards, public relations, education, lobbying, and arm-twisting ever seen to bring about even a semblance of an OCS leasing program outside the western and central Gulf of Mexico in the 1990s. More and more, for many companies, the likely result simply will not be worth the effort.

Lower 48 Onshore

· ·

Access to federal lands is a crucial issue for explorationists in the Lower 48 onshore as well, even if the prospects don't match the dazzling potential of ANWR's Coastal Plain and Offshore California.

Much of the debate revolves around the concept of wilderness preservation. There are 474 units totaling almost 91 million ac in the National Wilderness Preservation System. Outside Alaska, the western states hold about one-third of that total, with most of the acreage located in California at almost 6 million.

Under the 25-year-old Wilderness Act, Congress has the power to designate certain roadless scenic areas as wilderness, and therefore prohibited from multiple use management by federal agencies. The fight between environmentalists and industry should accelerate in the 1990s as the concept of wilderness takes on more of a quasireligious tone in the new consciousness. The fight is simple: Industry thinks too much land has been withdrawn already; environmentalists want as much as two to three times as has already been withdrawn added to the wilderness shelf.

Yet some in the environmentalist lobby itself have coexisted quite amicably with oil operations in their own little piece of the wild. For example, the Audubon Society owns the Paul J. Rainey Wildlife Sanctuary on the coast of Louisiana. In the 1970s, the society issued leases, created strict environmental stipulations, and still collects royalty checks averaging about $1 million/yr. The society's ownership of that land gave it the incentive to cooperate with an oil firm to find common ground for them both.

The same story can be told in the Big South Fork National River and Recreation Area in Kentucky's Cumberland basin, Big Cypress National Preserve in Florida, Rockefeller State Wildlife Refuge and Game Preserve in Louisiana, Aransas National Wildlife Refuge (site of the whooping crane's remarkable recovery from the edge of extinction) in Texas, and Theodore Roosevelt National Park in North Dakota. Each has oil and/or gas production, and each is a federally protected area. The distinction is that each is not also designated wilderness. They have maintained wilderness values and environmentally sound oil industry operations at the same time: multiple use of resources that accommodates concerns of both sides. Designating a tract of land wilderness eliminates the possibility for that partnership. An interesting test of that concept will be the Supreme Court's eventual ruling (accepted in January 1990) on the case of Lujan vs. National Wildlife Federation. That case questions the standing of environmental groups to challenge in court federal decisions to open public lands for development. If they suffer a defeat here, the environmental groups may pursue more such partnerships with industry, and it would be industry's best interest to accommodate them.

There are, however, other obstacles to multiple use on federal lands. The Forest Service introduced in January 1989 a new leasing system for national forests that could create new hurdles for operators of oil and gas leases. It includes a procedure for regional forest officers to make a suitability determination for land up for prospective leasing. In other words, a Forest Service office could determine an oil prospect was unsuitable for leasing by fiat—presumably if the environmentalists complained loudly enough.

There are other permitting steps in the Forest Service system—the right of the Secretary of Agriculture to halt operations on a lease if he deems a need for further environmental study, a requirement for surface use plans prior to seeking a drilling permit, and a bonding requirement stipulating 100%

of expected land reclamation costs—that are intended to comply with the National Environmental Policy Act. In other words, this leasing system does not clearly confer the right to explore for and produce oil and gas—so what is the use of buying a lease?

Two-stage leasing will become the prime battleground between environmentalists and Lower 48 onshore explorationists in the 1990s. The U.S. Supreme Court did nothing to clear up the confusion about whether oil and gas leasing triggers requirements under NEPA for environmental impact statements (EIS). The court essentially said review of a lower court decision holding that leasing does require an EIS is premature.

It will be left ultimately to Congress and the courts to determine piecemeal at what point the need occurs for an EIS—and thus the ripest opportunity for further stalling tactics by environmentalists to halt exploration. The likeliest outcome, given the direction of political winds in the United States today, is for more environmental study. Thus a two-tiered approach, with some sop to industry in the form of language governing no undue delays and nitpicking specifics over potential impacts after leasing, seems inevitable.

Is there a solution to the question of access to federal lands in the United States for the explorationist? Aside from all the short-term approaches, public education, and lobbying, it might behoove industry to press the federal government to take another look at revenue sharing, even if industry must surrender more of its share of revenues from federal lands oil and gas production.

Whether it involves ANWR's Coastal Plain, the Outer Continental Shelf, or the Lower 48, industry and the federal government must recognize the need to bring the public, in the form of state and local governments, as well as environmental groups, into not just the decision-making process, but also into the money-making process. This does not constitute buying off environmental concerns. It just means that, just as everyone who participates in an exploration venture assumes a certain amount of risk to earn a potential reward, it follows that those assuming a risk to another resource—the public's stake in a clean environment—also are entitled to earn a potential reward. It may prove explorationists' only real chance for access to America's petroleum resources in the 1990s.

8

DRILLING-PRODUCTION STRATEGIES

Nightmare off California

• •

Having gained access to federal lands for exploration and drilled discoveries of huge productive potential still is no guarantee of success.

Certainly not off California's golden shores. Just ask Exxon and Chevron, who each have seen the better part of a decade dissolve before being able to bring on stream the biggest oil strikes ever made in federal waters off the United States.

This has been the case for both companies despite the fact that there has not been a noteworthy loss of oil from drilling and production operations on the OCS since the 1969 Santa Barbara blowout. This has been the case despite years of review, permit approvals, regulatory oversight, countless environmental studies, and the most strenuous effort to mitigate environmental effects ever seen outside Alaska.

Exxon's saga of ensnarled development in the Santa Barbara Channel began shortly after the 1969 channel spill. The company discovered several large oil fields in the western portion of the channel in the late 1960s and with partners established the Santa Ynez Unit, encompassing primarily Hondo, Pescado, and Sacate fields. Reserves in the unit totaled more than 300 million bbl of oil. In the 1970s, Exxon began efforts to develop the first of those fields, Hondo (Spanish for "deep"), in almost 1,000 ft of water.

Although Exxon drilled the Hondo discovery well in summer of 1969, the final environmental impact statement and plan of development for the field were not approved until 1974.

From the beginning, the project was swathed in controversy. With the 1969 spill still fresh in mind, it quickly became apparent to Exxon that it would face fierce opposition in its plan to develop Hondo field as it wanted: to move the crude produced offshore to an onshore facility for treating, then to a marine terminal for tankering to market in Texas. Because of the low gravity and high sulfur and metals content of Hondo crude, Exxon had no West Coast refinery equipped to process it and instead had to ship Hondo production to its Texas Gulf Coast refinery.

The prospect of another major platform development project and increased tanker traffic off Santa Barbara County stirred up a hornet's nest of opposition spanning a decade of permitting snarls, regulatory delays, and lawsuits.

Although Exxon even won a county-wide referendum approving its development plan, continuing opposition led it to devise a fallback proposal that called for installing a converted tanker near the Hondo platform that would serve as an offshore storage and treating (OST) vessel. If stymied in its efforts to treat the crude onshore through denial of state or county permits, the company could elect to treat the crude, store it, and offload it into tankers outside the three-mile limit of state waters.

That, of course, is exactly what happened. After a decade, Hondo A platform went on stream in 1980, initially producing about 40,000 B/D into the OST for tanker shipment to Exxon's Baytown, Texas, refinery. Exxon's project survived only because the company took it outside the decision-making process that was killing it.

When Exxon sought to develop the western portion of Hondo field as well as Pescado and Sacate fields in the western half of the Santa Ynez Unit, it was confronted with the same kind of opposition. But this time the company had a trump card: going offshore. If still blocked in its efforts to further develop Santa Ynez Unit, Exxon had expansion of the OST as a fallback option again.

Santa Barbara County and the state place as a top priority avoidance of increased tanker traffic in the channel. Both have insisted that new production from Offshore California be moved to market via onshore pipeline if feasible. And the oil companies have been just as insistent that they have the flexibility to tanker in case a pipeline is unavailable or uneconomic or if they cannot ship to the refineries of their choice because of downstream environmental permitting concerns.

Basically, the companies and the county reached agreement that oil producers are obliged to prove that shipping via onshore pipeline is infeasible before being allowed to tanker their offshore production. With that finding of infeasibility, an offshore producer has the right to tanker until a feasible pipeline option is made available.

That situation has contributed to the impetus for construction of the All-American Pipeline (AAPL) system from Santa Barbara County to McCamey, Texas. The 1,200-mi, 30-in. pipeline moves about 90,000 B/D of oil from the San Joaquin Valley to transshipment points near McCamey.

Eventually, AAPL will be extended to the Gulf Coast area of Texas to link directly with the mammoth complex of refineries along the coast. A spur to the California coast has been completed, but the only crude that moves east is from the San Joaquin Valley.

When the rest of Santa Ynez Unit's platforms are installed and ready to start up, Exxon's production will move via AAPL to Texas. Exxon plans to complete installation of the platforms for developing the western portion of Hondo field and Sacate field in 1990 and is targeting a 1992 start-up for production that would peak at about 90,000 B/D in the 1990s. It also has begun construction of a 140,000 B/D onshore oil and gas processing plant at Las Flores Canyon near the Santa Barbara County coast.

To gain approvals for its further development of Santa Ynez Unit, Exxon worked out a compromise with the county that called for a package of emission reductions and tradeoffs that brought it back to its original plan of onshore treating.

Now Exxon plans to remove the OST in 1992; defer the unit's fourth platform until after peak air emissions from the second and third platforms are passed or more offsets are available; electrify key offshore equipment, including platform gas compressors, logging units, and crane; expand the onshore plant's cogeneration unit to 49,000 kw from 25,000 kw; use fewer supply boat trips with cleaner vessels; and pay for still more air quality studies.

In exchange, the county will allow Exxon to trade its OST emissions for some of Chevron's Point Arguello platform emissions offsets, obtain some of Chevron's onshore emissions credits for the Las Flores plant, and count as an offset credit the removal of the OST. At the same time, Exxon still will pursue construction of a larger marine terminal if it can prove to the county that shipping crude to the Gulf Coast via tanker versus a pipeline is economically justified.

However, nothing is certain regarding oil development in California. That can be seen in the labyrinthine regulatory struggle Chevron has undertaken in its efforts to develop Point Arguello field, at more than 300 million bbl of oil reserves the biggest oil field ever found on the U.S. OCS. Here again, the Exxon Valdez proved a turning point.

The drilling successes by Chevron, Texaco, Phillips, and other partners in the Point Arguello area have yet to translate into financial return. The discovery wells were drilled in 1980–83 in the area. Chevron and Texaco submitted formal development plans in 1984.

Those early plans called for production from Platforms Harvest and Hermosa to start up in November 1987 and Platform Hidalgo to go on stream in first quarter 1988. The three platforms were expected to be producing at peak as much as 90,000–100,000 B/D during 1990–93.

In addition, Point Arguello development was to serve as the infrastructure for a series of other developments of early 1980s discoveries in the area. Early plans called for Chevron to develop Rocky Point oil field, immediately east of Arguello, with a platform due to begin production in 1992. And further down the road, Chevron wants to develop Bonito oil field north of Arguello.

Chevron's early efforts on securing the necessary permits for Arguello development could stand as a textbook case on accommodating local communities' and government agencies' concerns on environmental issues. From the beginning, the company sought to bring residents into the planning and decision-making processes on Arguello development.

Its approach was even criticized privately by some California industry executives as giving up too much to the regulators and residents, resulting in an overdesigned project (in terms of environmental protection) that set an extremely difficult standard for other companies to measure up to.

We now know that that last opinion has proven far off the mark: It is impossible to give too much away in concessions to California's environmental extremists, and the Arguello story is a grim case in point.

Chevron hacked its way through a tortuous jungle of red tape toward Arguello start-up. It was stymied when Santa Barbara County seized upon the company's permit for levels of hydrogen sulfide in the associated gas to be shipped to the onshore treating plant.

Since the original plan had been submitted, subsequent drilling results indicated that H_2S levels would be much higher than predicted in the initial development plan. The county subsequently pulled its permit, and the ensuing dispute left the three platforms idle for two more years. Finally, a compromise was worked out between Chevron and the county that allowed the company to start with a more restrained, phased production approach that would limit the initial levels of H$_2$S. Chevron also agreed to provide a more extensive safety system for the pipeline carrying the H$_2$S-laden gas to shore.

After spending in all more than $250 million of the Point Arguello project's total $2.5 billion price tag on environmental mitigation measures, Chevron and partners were finally able to receive a permit from Santa Barbara County in early 1989 to proceed with start-up of field production—or so they thought.

GOO and the League of Women Voters appealed to the California Coastal Commission (CCC) to overturn Santa Barbara County's permit.

Unfortunately for Chevron and partners, Point Arguello became a victim of the Exxon Valdez backwash that was used to buoy the political fortunes of two of the commission members angling for higher office by posturing their respective ecological purity quotients before the public. The commission voided the county's permit, and the Chevron group was back at square one.

This time, however, Chevron was not quite so conciliatory. The company refiled its permit application for interim tankering and with its partners sued the CCC over the permit rescission.

Compounding the group's problems is a new controversy involving transportation of Arguello crude. A study that Arthur D. Little Inc. performed for the Arguello partners—part of their obligation under agreements with the county—studied the transportation options available to the Chevron group for moving Arguello crude to Los Angeles.

It found that the only feasible pipeline alternative available in less than three years was AAPL. And shipping Arguello crude via AAPL to the Gulf Coast carries a penalty of $4.25/bbl compared with the economics of tankering, the Little study concluded.

AAPL owner Celeron Corp. and Four Corners Pipe Line Co. disputed the Little study results and said that AAPL and Four Corners could transport Arguello crude to Los Angeles within a few months of approval and at a cost competitive with tankering—mainly through a spur to a Four Corners station in Kern County for blending at a nearby Celeron station. The pipeline companies also invoked the specter of the greater risk of an oil spill associated with tankering versus pipelines.

But Chevron countered the AAPL/Four Corners proposal by claiming that blending a condensate diluent to make the Arguello crude pumpable would add 50¢–$2.50/bbl to the proposed transportation charge. Celeron put the added charge at 5¢–10¢/bbl.

Chevron also noted that the diluent probably would have to be imported into Los Angeles on foreign tankers and then moved from Los Angeles to Kern County via more than 100 truckloads/day.

And, Chevron claimed, tankering would still be needed because AAPL at the most could take only 60,000 B/D of oil even with the pipelines' proposed changes, falling short of the field's targeted output of 100,000 B/D.

The claims and counterclaims arose anew with a revised study by Little early in 1990. That later study trimmed the gap between tankering and pipelining Arguello crude, but Chevron stuck to its guns, Celeron still criticized the later study, and Santa Barbara County seemed to be searching for confirmation of its contention that pipelining Arguello crude is economically and technically feasible for the Chevron group.

This standoff has produced nothing but frustration and added costs of $500,000 day in lost interest for Chevron and partners. But there may be a glimmer of hope ahead. Chevron's partners and shareholders are likely to step up pressure on Chevron to do something to end the impasse.

Meanwhile, the federal government in turn is likely to step up pressure on California and the local municipalities to ease up efforts aimed at blocking development of existing discoveries as a bargaining chip in exchange for a scaledown or halt to new OCS lease sales. With the Bush administration's virtual elimination of leasing off California this century, it is time for California to reciprocate on development.

Although the Arguello partners disdain the proposals put forth by Celeron and Four Corners, they must recognize the impossibility of developing a new pipeline route to Los Angeles. They also are not too keen on a crude exchange program—swapping Arguello crude sent to Texas via AAPL to take on another crude at Los Angeles.

So it seems that the best alternative might be some sort of accommodation that allows Chevron interim tankering until it can complete a feasible onshore pipeline to Los Angeles. That could only entail a new heated line along existing rights-of-way, probably Celeron/Four Corners'.

There are other pipeline right-of-way candidates, such as Mobil's line from the San Joaquin Valley to Los Angeles—identified as the least-cost pipeline alternative. But Mobil has its own agenda, planning to double San Joaquin Valley heavy oil flow to 100,000 B/D in tandem with a costly revamp at its Torrance refinery.

So a likely compromise would involve Chevron and partners agreeing to use double-hulled tankers, along with other tanker safety measures, while awaiting completion of a heated line along the Celeron-Four Corners corridors. There might still be concerns over nitrogen oxide emissions from the line's gas-fired heaters, but that probably could be remedied as a tradeoff on tanker emissions.

Such an accommodation could be presented as an effort to reduce tanker traffic off California's coast. Certainly the ultimate objective of pipelining Arguello crude to Los Angeles would be preferred by the state to continued tanker imports of Alaskan and eventually, foreign crude. That kind of scenario, given sufficient lead time on permitting, negotiating a memorandum of understanding, and lead time for tanker construction, could mean a start-up for Arguello before 1992.

The dispute promises to continue through 1990–91, fueled more by Earth Day hype as campaigning California politicians seek to outdo each other in environmental purity by pressing even more stringent oil spill and tanker safety measures.

The only recourse left to Chevron and partners and others seeking to develop oil fields off California then may be that which faced Exxon: either find a way to make economic use of a pipeline—which may not be possible; walk away from the investment—perhaps unthinkable for the Arguello group; or go Exxon's route—all processing and tanker loading offshore.

Taking all processing offshore for the Arguello group would be a much more difficult task for Chevron than it was for Exxon. Exxon's OST is designed for 40,000 B/D from Hondo field. A similar arrangement at Arguello would entail a much bigger OST and either more frequent tanker loadings or bigger tankers.

Even though Exxon ostensibly has prevailed through the leverage of its fallback option of installing a larger OST layout involving increased tanker traffic, it accomplished that at a time when the legal and permitting climate was not so imbued with the Holy Crusade of Earth Day 2. Still, it may be the only approach available for further development plans off California, if opposition to even interim tankering and any onshore pipeline projects continues.

Although it appears that further development of the Santa Ynez Unit will proceed, it would be premature to assume that no further roadblocks lie ahead for the project. Given the climate today in California, it would not be surprising to see more efforts by environmental groups, regulatory authorities, and politicians seeking to halt Exxon's project again, probably through the courts.

The litigious weapon of choice for opponents of development probably will continue to be the CZMA. CZMA provides a means for states and local communities to provide input on any activity, notably resource development, that affects coastal regions.

The key to CZMA is "consistency"—that activity be consistent with coastal zone management plans set by states. California has led the way in utilizing CZMA as a veto over any activity in federal waters it did not like.

Although unsuccessful in the courts to halt oil and gas leasing on the OCS through CZMA, the state, through the CCC, has repeatedly blocked drilling and production offshore and along the coast by invoking CZMA powers. The only recourse available to companies whose projects have been stymied by CCC actions is to appeal to the U.S. Secretary of Commerce. When CCC bans have been overturned by Commerce, the agency or environmental groups turn to the courts again.

Even now, there are congressional efforts to toughen the consistency provision of CZMA. A House merchant marine oceanography subcommittee approved legislation in spring 1990 reauthorizing CZMA, including provisions that any federal activity, regardless of its location, must be consistent with state-approved programs in the coastal zone. Essentially, the provision is intended to counter the effects of a 1984 Supreme Court ruling that found that OCS lease sales are not subject to consistency requirements. Under the provision, the President could exempt a federal agency's activity from the consistency requirement if it is in the paramount interests of the United States.

The reauthorization bill also seeks to establish a Coastal Zone Management Service to be the lead agency for coastal zone management programs.

One industry lawsuit seeks to blunt the drive of states and local communities to intervene more directly in the process of permitting development projects off California's coast. A suit by the Western Oil & Gas Association (now WSPA) and the National Ocean Industries Association seeks to have declared unconstitutional the recent flood of California municipal and county ordinances that forbid the presence of onshore facilities, such as pipelines or processing plants, related to offshore development projects.

Shell's plans to develop San Miguel field off San Luis Obispo were hobbled by such local ordinances. The city and county of San Luis Obispo approved local rules calling for any onshore activity related to offshore development to be put to a referendum among voters.

Recognizing the slim chance of such approval, Shell has shelved its permit application to bring San Miguel crude ashore at San Luis Obispo and now seeks to move the oil through a subsea pipeline to landfall at Santa Barbara County. That may seem like going from the frying pan into the fire, but if a condition of development is that industry must utilize a single, consolidated onshore processing facility, it may well prove to be the Shell group's only chance for developing the field.

Another court suit over an Offshore California oil project could have far-reaching implications for the petroleum industry.

In 1987, ARCO sued Santa Barbara County and the State Lands Commission (SLC) over a May 27, 1987, SLC decision denying ARCO a permit to develop Coal Oil Point field in state waters of the Santa Barbara Channel. The project entailed installing three platforms to produce at peak 80,000 B/D of oil and 150 MMCFD of gas.

ARCO is taking the novel approach that could prove a landmark for the industry, with broad implications in other operations. ARCO wants the courts to force SLC and the county to approve its development plan or pay damages of about $840 million. The thrust of its argument is that if an agency has the power to deny approval of an economically feasible project, such action is a taking of property and requires just compensation to the owner.

That implies, for the first time, that a regulatory body's action blocking oil and gas work carries a big price for that body. ARCO was unsuccessful in Los Angeles Superior Court in early 1990, but its appeals will bring it before friendlier courts in the future.

There are also lawsuits over the issue of regulating air emissions from oil work off California. Efforts to have authority for regulating OCS emissions transferred to local jurisdictions from Interior may well prove moot. After President Bush's directive—part of his lease sale announcement in summer 1990—new emissions regulations from Interior's MMS covering OCS activity will mirror California's own stringent rules.

There are many other projects, from single exploratory wells to sizable development programs, that remain blocked by local and state actions. Aside from the legal action, the lobbying, the negotiating, and the accommodations, what other options are there for oil and gas companies to develop discoveries estimated conservatively in the range of several billion barrels of oil reserves off California?

One solution that has often been bruited about but never received the serious attention it deserved is a change in the split of 8g money, the funds the federal government collects from OCS rents and royalties and returns to the coastal states.

As it stands now, California, Alaska, Texas, Louisiana, and Alabama receive

27% of the royalties and lease payments from OCS activity within 10 mi of state waters, which end at the three-mile line. Outside the 10 mi limit, the federal government gets 100%.

By contrast, the federal government shares 50–50 revenues it receives from onshore mineral development within states where that occurs, except Alaska, which receives 90%.

The Bush administration probably will pursue the idea of increasing the states' take of 8g money. There is the obvious benefit of directing more of the offshore revenues to the regions that are directly affected by offshore activity, thereby helping to mitigate those effects.

Some communities opposing offshore development have charged that such a move amounts to an attempt to buy off coastal communities in order to expedite permitting. Even with their opposition, however, there will be enough communities that need and will welcome those increased revenues. That could create a split in the ranks of coastal communities, diluting some of the opposition.

And all communities, once a new or expanded revenue stream is introduced into budgets, tend to expand their needs to become reliant on those added revenue streams. Alaska is a perfect case in point. Except where politically untenable, such as Bristol Bay with its powerful fishing interests, Alaska generally has supported development on its federal lands, onshore and off. Being 85–90% dependent on oil revenues no doubt underpins that consistency of support.

The Laguna Beaches of the world may turn their noses up at a few extra million dollars per year, but there are tenfold as many communities along the Pacific, Atlantic, and Gulf coasts that could certainly find a use for a bigger chunk of a revenue source that by midyear 1989 had provided more than $90 billion. Once established, even the Laguna Beaches may well find a regular use for those increased revenues.

Alaskan Dilemmas

· ·

Alaskan oil and gas operators face problems in drilling and production activities beyond the issues of wetlands loss and causeways covered in the previous chapter.

DRILLING-PRODUCTION STRATEGIES

Regulations governing drilling and production operations on Alaska's North Slope, perhaps the most costly and tightly regulated in the world, will become even more stringent as opponents of ANWR leasing target industry's record there to justify putting ANWR off-limits.

One way the environmentalists are directing the debate is by shifting the focus from actual impacts to the ecosystem—for which they have been unsuccessful in finding a "smoking gun"—to playing a numbers game on permit compliance.

That is because industry has concentrated on wildlife studies to gauge the effects of operations on the environment and thereby grade their environmental record. Because the federal agencies monitoring that record have found that the environmental impacts have been pretty much in accord with what was expected, focusing on permit exceedances over a long period of time—including periods reflecting abandoned practices—tend to suggest a pattern of abuse.

For example, ARCO was sued in 1988 for discharging or allowing leakage of water that collects in its 200 North Slope reserve pits for discarded drilling muds and cuttings onto the tundra.

Although it had a state permit to dewater the pits after meeting state water standards, and EPA deemed that sufficient, the company later realized it should have been more aggressive in obtaining a federal permit.

While the state drafted new regulations, ARCO filed a new plan for handling the reserve pits. It stopped dewatering—instead removing the water from the pits, hauling it off, and disposing of it—pending approval of a three-year, $50-million plan to close out or line the pits.

In future, ARCO plans to bury cuttings in nearby holes where they would be permanently frozen. But environmentalists want to see the drilling wastes minimized, segregated, detoxified, and recycled.

The point here is not whether occasional permit exceedances contributed to ecological damage. Although the cuttings and muds may be harmless and the reserve pit water drinking quality, the practice gives the environmentalists ammunition. They can cite such red flag phrases as "permit violations," "toxic metals," and "contamination" to raise the specter of ecological harm without having to prove or even make a strong case that these practices are indeed causing damage.

North Slope operators are responding by ensuring that their operations do everything possible to minimize, reclaim, recycle, and generally micromanage waste from drilling-production operations.

Scientifically sound or not, the regulatory thrust on drilling-production waste on Alaska's North Slope and offshore is moving toward a zero-risk standard. That should be kept in mind by operators, contactors, and service/supply companies as they continue to fund studies of what is and is not ecologically harmful. The presumption of innocence has little place here.

Beyond the issue of "no net loss," new guidelines for the protection and management of federal wetlands also pose a threat to North Slope development. In April 1989, EPA and the Army Corps of Engineers published for public comment draft guidance identifying most of the Colville River delta as unsuitable for disposal of gravel fill material because it is a sensitive bird habitat.

The area in question is a critical one because of several noncommercial heavy oil discoveries by Texaco and Amerada Hess there, and several other companies have leases there. If the guidelines proposed are adopted, the companies would be denied the right to develop those or any future discoveries.

Waste Issues

The issue of regulating the handling and disposal of drilling-production wastes is one that many operators might think has been largely defused in the 1990s.

It is easy to be lulled into thinking that, since most drilling-production wastes are exempt from the Resource Conservation and Recovery Act (RCRA) Subtitle C that covers disposal of hazardous substances.

Substances covered by Subtitle C include used lube oils, unused frac fluids or acids, used hydraulic fluids, paint waste, radioactive tracer waste, and waste solvents.

Most drilling and production wastes are classified as nonhazardous under RCRA Subtitle D, less stringent guidelines mainly under state control. They include produced water, drill muds and cuttings, reserve pit waste, and rig washdown water.

Especially critical to staying out of the tougher Subtitle C guidelines is that an operator must avoid mixing wastes designated hazardous with nonhazardous wastes. Even traces of a nonhazardous substance can lead to a designation of hazardous for a much larger volume of otherwise nonhazardous waste.

It isn't just a question of more costly handling and disposal methods when dealing with hazardous waste. There are a number of regulatory arenas that come into play when substance is deemed hazardous. They include:

- SARA Title III (Superfund Amendment and Reauthorization Act of 1986), which establishes a system of reporting to local communities to keep them abreast of hazardous substances used or stored at a facility.
- OSHA's similar right-to-know guidelines under the Hazard Communications Act requiring that workers be informed of the hazards of any material at a facility.
- Superfund (Cercla) rules governing releases, cleanup, and disposal of hazardous materials.
- Department of Transportation rules regulating shipment of hazardous materials.

This exponential expansion of regulatory oversight that comes with the designation of hazardous carries with it an equally exponential expansion of costs.

So it follows that operators of drilling and production facilities should go to great lengths to eliminate mixing of wastes, reduce the amount of waste as much as possible, detoxify wastes, and guarantee their proper disposal.

Recycling and reconditioning of all wastes should be a priority—and can have their own economic benefits. If a substance is to be reused, it is not classified as a waste. Disposing of waste outside an operators's facility does not, however, eliminate his liability, even when the disposal facility disregards regulations.

What may escape casual scrutiny in the classification of hazardous versus nonhazardous substances is that the EPA's exemption of many drilling-production wastes from its hazardous list is *temporary*.

In addition, there are sufficient ambiguities in how EPA classifies certain wastes, using such qualifiers as "possible" or "believed to be," that an operator cannot truly be certain that a substance he thinks is exempt today might not later fall under the cradle-to-grave purview of RCRA Superfund, and the panoply of other hazardous waste laws.

Given the status of law over orphan dump sites and third-party liability, the operative phrase here might be "better safe than sorry."

In the case of wastewater from drilling-production, the regulatory ambiguity stems from regulations that are still being formulated.

EPA is formulating regulations under the Clean Water Act covering discharge of drilling fluids and produced water effluent. The regulations will fall under one of the following categories: best practicable control technology currently available (BPT), best available technology economically achievable (BAT), best conventional pollutant control technology (BCT), and new source performance standards (NSPS).

BPT, which forbids discharge of free oil, has been in effect since 1977. EPA proposed the other guidelines for certain waste streams from offshore facilities. Other waste streams will be covered by a new proposal likely in 1990, with a final rule to be implemented in 1992.

For coastal operations, EPA will implement BAT, BCT, and NSPS guidelines—to join BPT rules in effect now—by 1995.

What will form the critical test of these new guidelines is the determination of toxicity of contaminants in the effluent stream. Statistical studies of bioassays conducted in support of clean water discharge permits are in dispute.

It would be well worth industry's investment to fund further biological studies to help determine a useful guideline for toxicity of effluent streams from drilling and production operations.

This issue won't just affect offshore and coastal operators. EPA now is looking at effluent discharge rules for onshore stripper wells. That would expand this area of regulatory oversight to as many as 450,000 wells, the vast majority of those operating in the United States.

Industry needs to provide as much input as possible into EPA's rulemaking process on what kinds of mitigation efforts are needed.

Air Quality Concerns

· ·

The U.S. upstream petroleum industry may have dodged a big bullet when the Senate voted in spring 1990 to amend its CAA reauthorization bill to exempt emissions from oil and gas production facilities as air toxics subject to regulation. Estimates were that the cost of such added controls would run to more than $9 billion, and thus the United States would have lost as many as 25% of its producing wells had they been subject to those standards.

But industry is not out of firing range yet. In areas of ozone nonattainment, the application of ozone controls to sources of less than 100tons/yr of pollutants would be a real threat to oil and gas operations. EPA is considering added controls on emissions of nitrogen oxides versus its previous emphasis on controls of volatile organic compounds. VOCs and NOx are among the chief precursors of atmospheric ozone.

Ironically, the mechanics of how they react in sunlight to form smog is complex and little-understood. Some studies suggest that, depending on certain conditions and the ratio of VOC to NOx, cutting NOx emissions actually can increase ozone levels instead of reducing them.

Under some proposed ozone bills, NOx controls would be mandatory, and the threshold at which a facility would be considered a major stationary source would be slashed to as little as 10–25 tons/yr. Under a 10 tons/yr definition, NOx controls would be mandated on a 100 horsepower engine.

Another major concern on ozone is the consideration of expanding ozone nonattainment areas on the theory that ozone and its precursors are wind-blown from one area to another; thus pollution in a noncontrolled area could aggravate conditions in a nonattainment area. That could call for expanding nonattainment areas to entire metropolitan areas versus counties, and with that, expansion of tighter controls.

Operators of drilling and production facilities must contend with the growing entanglement of emissions offsets—reducing air emissions elsewhere in a nonattainment area in order to obtain a permit for a new facility or modification to an existing facility that would be a new source of emissions.

The offsets in an area like California could be as much as a ratio of 4.1-to-1 and are at a minimum of 1.1-to-1. As with almost every other environmental trend that originates in California, look for these tougher offset ratios to crop up in other nonattainment areas.

Drillers and producers also need to keep the debate focused on net air quality benefits, not just predetermined annual cuts of emissions, until attainment is achieved. Federal air quality policy has resulted in sharp reductions of emissions per ton over the years without a net benefit in air quality in many instances.

Looming on the horizon is the threat of new global warming mandates. Carbon dioxide, methane, and NOx have been identified as purported culprits in this debate. Now that the South Coast Air Quality Management District has seen fit to take responsibility for Southern California's contribution to the greenhouse effect as well as depletion of the ozone layer and smog (what's next—purview over television broadcast signals emanating to the stars?), look for the first controls affecting emissions of so-called greenhouse gases from drilling and production operations to be applied in the Los Angeles basin.

There will no doubt be tighter controls attempted throughout the nation, and certainly in nonattainment areas, of regulations governing fugitive emissions from oil and gas field facilities.

California is implementing tough new rules amending existing "sniffer" rules for inspecting fugitive emissions from valves, pumps, and other oil field components. The California Independent Petroleum Association estimated that new sniffer rules imposed by the Scaqmd would cost many producers at least $4,400/yr per lease for inspections. That works out to a staggering $419,000/ton to remove those emissions, CIPA estimated.

Although that far exceeds the agency's own cost-benefit standards, Scaqmd had no trouble approving the new rule. Scaqmd earlier approved similar fugitive emissions controls for wastewater separators and production sumps.

In addition to air quality concerns posed by oil field facilities, producers need also to be aware of the threat to wildlife posed by uncovered oil pits and storage tanks. According to the U.S. Fish & Wildlife Service, more than 100,000 waterfowl die each year after landing in these uncovered oil receptacles, thinking they are bodies of water.

The agency will accelerate its enforcement of rules governing protection of waterfowl and other migratory birds in the 1990s. Operators convicted of causing the deaths of these birds could face penalties of six months in jail and $10,000 fines. In the case of endangered species such as the bald eagle, the punishment could be $210,000 in company fines, $300,000 in individual fines, and as much as two and a half years in jail.

Considering the damage to a company's reputation (remember the still-fresh images of the oil-soaked wildlife in Prince William Sound), not to mention the penalties, the cost of netting these facilities seems trivial indeed.

Drilling Safety

• •

The revolution occurring in the industry governing drilling rig safety predates the revival of environmental consciousness. It can be traced directly to the industry's worst accident, the massive explosion and fire that claimed the Piper Alpha platform and 167 lives in the U.K. North Sea in July 1988.

There has been much debate over whether the Piper Alpha tragedy and the subsequent string of offshore drilling and production accidents, especially in the North Sea, are linked to industry budget cuts in the wake of the 1986 oil price collapse.

Whether or not that contributed to recent mishaps, there is one thing certain: Drilling rig designs and procedures will continue to focus on safety considerations as well as thriving in a low-cost environment.

Operators will emphasize reduced exposure to hazardous materials by increased automation of drilling equipment. Furthermore, they will improve use of computers to enhance well control technology. And operators will develop more advanced systems for handling gas.

Companies will have to invest more in research of automation and robotics in equipment and materials handling on rigs not only to reduce risk to personnel, but also to reduce risk of accidents threatening the facility and the environment. Special targets for automation in rigs in the years to come are pipe and mud handling systems.

Despite advances in well control technology and the information systems that accompany them, loss of rigs due to blow-outs increased during the 1980s, according to Offshore Data Services Inc., Houston.

Most of those incidents have involved shallow gas zones where diverter systems have failed, according to Veritec, Oslo. Eventually, more and more drilling operations will come under the control of information management systems.

Safety oversight should come into play for every new procedure and regulation implemented for a drilling rig. It stands to reason that introduction of a new procedure will be accompanied by an increase in accidents if training is insufficient. Risk assessment must be pursued religiously in implementing any unfamiliar activity.

In safety considerations, of course, people are the paramount concern. Not only should they be the focus of protection, they should also be the focus of accident prevention.

According to International Association of Drilling Contractors data, the rate of lost-time accidents on drilling rigs fell by more than three-fourths in the past 27 years. That is the result of improvement in equipment and safety procedures, tougher regulations, improved working conditions and training, and expanded drug and alcohol screening.

However, recent IADC data suggest that trend has weakened since the 1986 price collapse. There is reason to believe that this deterioration is linked to the high turnover of experienced personnel after the price collapse. Apparently, improvements in training may not be compensating for the loss of this experience. So industry must redouble its efforts to ensure training and screening programs keep up with standards and try to create a working atmosphere that cultivates the buildup of a high level of experienced personnel in drilling operations.

A Last Word

Perhaps the best single environmental argument that U.S. oil and gas companies can make in favor of domestic drilling-production is that constraining U.S. drilling-production actually increases the risk to the environment.

Every barrel of oil produced in the United States is one that is not imported into the United States. Producing less U.S. oil means importing more via tanker and with it increasing the risk of an oil spill.

Further, the economic benefits of drilling and production in the United States spread far beyond the oil patch states. Various estimates put the job-creation benefits and economic ripple effects of drilling and production at anywhere from 2-to-1 to 4-to-1 in comparison with the number of jobs the activity creates directly.

So far, opponents of U.S. drilling and production activity in sensitive if highly prospective regions have been successful in linking that activity in the public's mind to oil spills such as those off Alaska's coast in 1989 and California's coast in 1990.

The single most crucial challenge for the drilling-production industry in the environmental decade will be to use the truth as a weapon to destroy that false linkage.

And the surest way of undermining that effort will be a string of tragic accidents like those in the North Sea in recent years or a single massive spill from a blowout.

Don't presume such a blowout-and-spill cannot happen, no matter how unlikely it seems in light of industry's record since the 1969 Santa Barbara Channel spill.

208

9

TRANSPORTATION STRATEGIES

The Exxon Valdez spill has left the U.S. petroleum industry's transportation sector with perhaps the greatest urgency among oil and gas operations with regard to environmental mitigation measures.

As damaging as that event was to industry's image, the damage would have been substantially mitigated if the petroleum transportation industry had not evinced a propensity for shooting itself in the foot with distressing regularity since then. A string of transportation accidents since the Exxon Valdez through summer 1990's Mega Borg accident off Galveston, Texas—some of which resulted in spills and/or deaths and injuries—has kept the image of an exceedingly clumsy or cavalier approach to the business of transporting petroleum engraved into the public consciousness.

So it follows that there is no more compelling need to bolster the public's faith in the industry's commitment to environmental preservation than in the business of transporting oil and gas.

The Exxon Valdez spill itself was enough to resurrect the notion of transporting Alaskan crude via an overland route to the Midwest, which many Alaskans had favored in the first place. U.S. concerns over having such a vital supply line on foreign, albeit friendly soil, and oil companies' wishes to ship Alaskan crude to Japan junked the idea in the 1970s. Perhaps a tradeoff for exploring for and developing oil on ANWR's Coastal Plain would be an insistence on ending tanker traffic from Valdez. Will this idea fly? It seems to fly in the face of the reality of getting another TAPS-equivalent permitted and built across more wilderness terrain in two countries and the powerful opposition of the maritime industry (which was pretty effective in shutting down the Alaskan crude-to-Japan option).

TRANSPORTATION STRATEGIES

In any event, the key transportation issue is avoiding spills and leaks, especially from tankers. Forecasters are virtually unanimous in predicting that U.S. oil imports will rise sharply in the 1990s as demand continues to rise and domestic production continues to plummet. If follows then that there will be an increasing risk of catastrophic oil spills in U.S. coastal waters, right?

Not necessarily. Environmentalist groups used the first anniversary of the Exxon Valdez spill to decry the nation's growing dependence on oil use as a harbinger of ecological doom. The Wilderness Society that anniversary week issued a report that there had been more than 10,000 oil spills in the United States since the Exxon Valdez incident, spilling an additional 4–9 million gal. of oil. That figure includes pipeline as well as tanker spills, onshore as well as offshore.

What the Wilderness Society did not point out at its press conference touting the report is that if one subtracted the major spills—typically considered as 10,000 bbl of oil or more—that leaves a maximum average of only 320 bbl/spill, hardly a catastrophic measure.

In fact, the typical spill more appropriately measured in terms of single digits. As far as MMS is concerned, *any* leakage, even less than 1 barrel, even an oily sheen on the water of unknown origin, must be reported. So a passing fishing boat with a leaking motor or a little grease on a platform deck floor that is washed overboard by rainfall suddenly becomes a "spill." That can add up to 10,000 "spills" or more in a hurry, yet it is only a fraction of the amount of oil that actually ends up in U.S. waters.

According to the International Tanker Owners Federation (Intertanko), about 23.4 million bbl of crude and petroleum products leak into the world's oceans every year. Tanker accidents account for only 12% of that total, and offshore oil and gas exploration and production operations only 2%. Normal maritime operations (e.g., the leaking fishing boat motor) account for 33% of the offshore oil pollution, and industrial and municipal waste discharge accounts for 37%. The remainder comes from atmospheric fallout (7%) and other natural sources (9%).

Further, says Intertanko, there were only 10 major tanker oil spills in the 1980s, compared with 20 in the 1970s.

In fact, the U.S. tanker industry's record on oil spills even improved in 1989, the Exxon Valdez spill notwithstanding. There were fewer major oil spills in 1989 than the average during 1978–89, according to a study by *Golob's Oil Pollution Bulletin* of Cambridge, Massachusetts. Golob's used 10,000 gal.—

Total Number of Tanker Spills in U.S. Waters by Year: 1978-1989

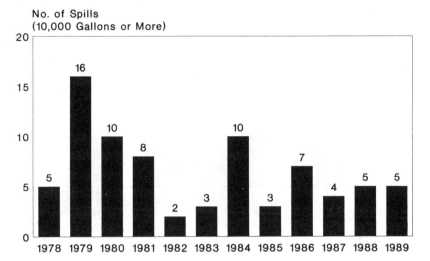

No. of Spills (10,000 Gallons or More)

Note: The figures for 1981 include 2 spills of undetermined size.

Source: *Golob's Oil Pollution Bulletin*

Total Amount of Oil Lost in Tanker Spills by Year: 1978-1989

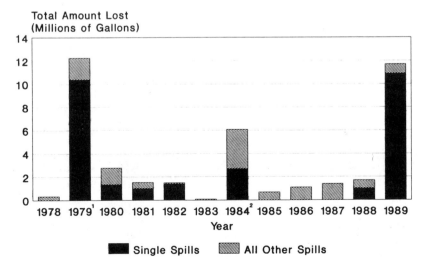

Total Amount Lost (Millions of Gallons)

Year

■ Single Spills   ▨ All Other Spills

[1] The Burmah Agate incident involved 7.8 million gallons burned and 2.6 million gallons spilled.

[2] The Puerto Rican incident involved 1.7 million gallons burned and spilled and 336,000 gallons sunk.

Source: *Golob's Oil Pollution Bulletin*

Figure 9–1 Tanker spills in United States waters for 1978–1989.

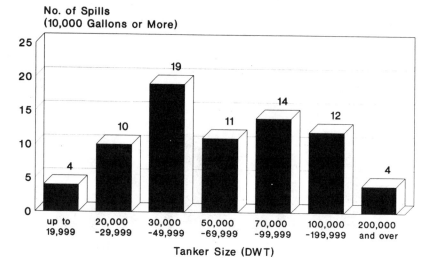

Tanker Spills in U.S. Waters by Vessel Size: 1978-1989

No. of Spills
(10,000 Gallons or More)

Tanker Size (DWT)

Note: The 6 spills from vessels of
unknown size were excluded.

Source: *Golob's Oil Pollution Bulletin*

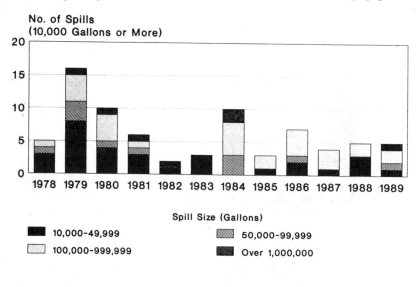

Tanker Spills in U.S. Waters by Spill Size and Year: 1978-1989

No. of Spills
(10,000 Gallons or More)

Spill Size (Gallons)

10,000-49,999   50,000-99,999
100,000-999,999   Over 1,000,000

Source: *Golob's Oil Pollution Bulletin*

Figure 9-1 (Continued)

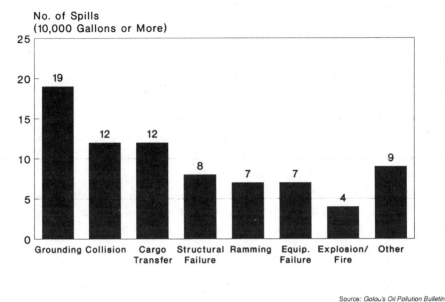

Causes of Tanker Spills in U.S. Waters by Year: 1978-1989

No. of Spills
(10,000 Gallons or More)

Source: *Golob's Oil Pollution Bulletin*

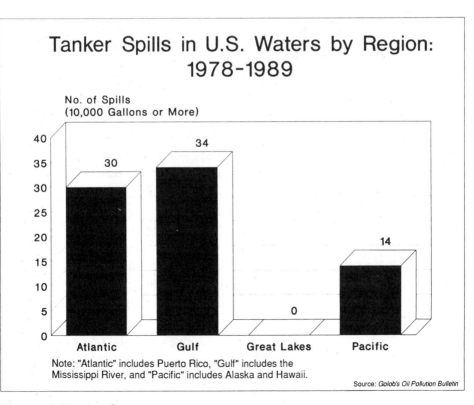

Tanker Spills in U.S. Waters by Region: 1978-1989

No. of Spills
(10,000 Gallons or More)

Note: "Atlantic" includes Puerto Rico, "Gulf" includes the Mississippi River, and "Pacific" includes Alaska and Hawaii.

Source: *Golob's Oil Pollution Bulletin*

Figure 9-1 (Continued)

not barrels—as its criterion for "major" (Fig. 9–1). There were only five tanker spills involving more than 10,000 gal. (about 238 bbl) in 1989, compared with an average of six or seven per year in 1978–89.

There may have been a correlation in the volume of tanker traffic in U.S. waters and the number and volume of oil spills. According to Golob's study, the peak year for the number of tanker oil spills covered in the study period, 1979, also was the peak year for total U.S. crude and products imports and in terms of oil volumes lost—not just spilled—from tankers in U.S. waters.

Total U.S. oil imports are rapidly approaching the volume levels of 1979 on an average annual basis. Will that mean more tanker oil spills? Was 1989 an anomaly in terms of the number of tanker spills?

There are too many variables to easily establish a link between tanker oil spills and the levels of tanker traffic in U.S. waters, says Richard Golob, publisher of the newsletter. The decline in the number of 1989 spills could be the result of tighter safeguards in the aftermath of the Exxon Valdez, simple luck, or both.

Common sense would suggest, however, that greater frequency of tanker trips means greater risk of spills. The leading proponent of that view outside of the environmental movement now is, ironically, the U.S. petroleum industry.

The API points out that banning offshore oil and gas exploration could worsen the risk of oil spills in U.S. waters, because every barrel of oil not produced in the United States is one that must be imported into the United States—and more than 75% of the oil imported into the United States comes via tanker, according to API.

The association makes this linkage to counter the environmentalists' linkage of increased offshore drilling with higher risk of oil spills. Unfortunately, it also plays into the environmentalists' hands when they can in turn counter that the real problem is U.S. dependency on *any* oil, which then bolsters support for their argument in favor of radical conservation and fuel efficiency measures.

The irony of this situation was highlighted when Sen. James McClure (R.-Idaho) introduced an amendment to the Interior Department appropriations bill in 1989. It would have banned oil tankers from federal waters that were also off limits to leasing under another amendment to the appropriations amendment. His reasoning: Areas that were so ecologically sensitive as to be free of drilling should not face the much greater risk of spills from tankers. McClure's proposal did not survive, but the idea is a telling one—and one that industry would be wise to repeat in the 1990s to illustrate the dilemma America faces in dealing with environmental and energy concerns—and the correspondingly unequal sharing of the burden for environmental risk among coastal states.

At the same time, the petroleum industry by its own admission now acknowledges that it does not have the capability to deal with an Exxon Valdez-scale spill in U.S. waters. Even with major capital outlays and significant advances in containment and cleanup technology, it will be logistically impossible to handle a spill of that size anywhere, at anytime. Marine mishaps happen, and they always will.

The crucial thing to remember today is that the public perceives the environmental and safety risks of transporting petroleum as getting worse, not better. Even if the number of major spills or total spills is less than in the 1970s, it may be that a petroleum industry faced with an aging pipeline infrastructure in the United States and an aging tanker fleet worldwide, together with personnel stretched far too thin, will see the number of total and major spills increasing in the next few years.

The appalling record at Arthur Kill in New York Harbor in 1990 may point to that. Even after Exxon undertook measures to improve safety of tanker loading and offloading at its Bayway, New Jersey, refinery and Linden, New Jersey, terminal in the harbor, the harbor suffered its sixth significant spill of the year as this chapter was being revised.

With that in mind, the petroleum industry must make tanker safety a top priority for today. It took a big step toward that with the creation in 1989 of the Petroleum Industry Response Organization, which is the centerpiece of a three-part industry oil spill response and cleanup program that will cost more than $250 million the first five years alone.

Tanker Routing Issues

. . . .

Senator McClure's somewhat tongue-in-cheek amendment aside, there are some serious proposals to ban or shift tanker traffic from sensitive areas.

Early in 1990, companies representing 25% of the tanker traffic through the Florida Straits agreed to move their tanker traffic 10 mi away from the Florida Keys. The move was intended to protect coral reef and other sensitive coastal resources from the threat of groundings and oil spills.

Chevron Corp. and Unocal Corp. two months after the Exxon Valdez spill agreed to reroute their tankers around the Santa Barbara Channel. The cost: an extra $500,000/yr because of the increased distances in tanker trips.

Some environmental groups want to go even further. The Center for Marine Conservation has called for a vessel free zone in the most ecologically sensitive areas, such as the Florida Keys and the central California coast. Under the CMC proposal, cargo ships would be prohibited from passing within 25 mi of the coast. They also would be limited to a 5-mi wide fairway in order to control vessel routing.

There is no doubt that stricter controls will be placed on tanker routes, in establishing sea lanes and adhering to them. The Alaska oil spill commission, issuing a report in early 1990 on the Exxon Valdez spill, noted that the vessel was running at full speed outside designated tanker lanes at the time of the accident. The commission said that rules about staying within the lanes at all times in Prince William Sound and slowing to minimum speed when ice was present were consistently violated and that the Coast Guard stopped enforcing the rules.

There is another way of dealing with the issue of tanker routing: deepwater mooring terminals. Interior Sec. Manuel Lujan gave that idea its widest audience when he committed his agency on national television to a study of requiring mooring terminals at least 5 mi offshore. By his reasoning, offloading tankers in deep water away from ports would decrease the risk of tanker accidents, eliminate the need for offloading at some ports, and in the event of a spill would allow more response time and thus cut the risk of spilled oil reaching shore.

A similar idea was put forth by attorney Joseph Petrillo, the former director of the California Coastal Conservancy. Petrillo maintains that an Exxon Valdez-scale spill is just waiting to happen in the Santa Barbara Channel. In addition to local traffic such as Chevron's and Unocal's tankers, the 30–40 mi wide channel is traversed by tankers from Alaska en route to the Long Beach and Los Angeles harbors and to the Trans-Panama Pipeline, passing closely by oil platforms. Petrillo proposes that the industry install a single monobuoy anywhere from the Santa Maria area to Point Conception along California's central coast to divert tanker traffic from the channel. Oil could be offloaded outside the channel and moved to shore via subsea pipeline, in Petrillo's vision, to link with the All-American Pipeline to Texas. It would also have the effect of consolidating several marine terminals in the channel.

Unfortunately, Petrillo misses a few key points. One is that AAPL has a capacity of only 300,000 B/D, and Exxon is committed to shipping its Santa Barbara Channel crude via AAPL—another 100,000 B/D in addition to the 100,000 B/D of onshore California crude moving through AAPL now. Further, the volume of Alaskan crude moving to Texas now totals more than 600,000 B/D. Another point is that some of the shippers of Alaskan crude—accounting for about another 400,000 B/D—want to land their crude at refineries in California, not in Texas. That resurrects the controversy over shipping crude from an offshore source (tanker or new platform) via onshore pipeline, discussed in the previous chapter.

The industry would be wise to be leery of such proposals, although more catastrophic spills might give these ideas added momentum. There is such a deepwater mooring terminal, and it has proven less than a rousing success. The Louisiana Offshore Oil Port (LOOP) was built in the mid-1970s after long delays, and has had economic difficulties from the beginning.

In a July 1990 report, MMS concluded that deepwater offshore oil terminals may help reduce the risk of nearshore tanker accidents and oil spills, but there remain significant questions about their economic and environmental effects.

MMS contends more study is needed before it could be determined that the reduced risk of oil shipping would justify the terminals' costs. It would take about 11 deepwater terminals to handle about 80% of the crude and products moving in and out of medium and large U.S. ports. The agency estimates construction costs would range from $69 million for a single buoy system to handle 400,000 B/D in a relatively sheltered environment in 125 ft of water to $1.5 billion for a multiple berth system to handle 1.6 million B/D in 120 ft of water 32 mi offshore.

Costs of pipelines associated with such terminals also jump sharply, according to the MMS report. The agency estimates terminal pipeline costs will range from $1.8 million/mi in the Gulf of Mexico to $2.3 million/mi in Prince William Sound.

Operating costs at LOOP in 1989 were about $256,000/day for the facility, with throughput at 800,000 B/D, a little more than half its capacity.

"A major reason for the lack of interest in constructing new offshore terminals has been LOOP's failure to return an investment for its owners," MMS said. "Therefore, without some form of government encouragement, offshore terminals will attract little interest from investors."

That "encouragement" could mean government subsidy or indirect monopoly creation. If tankers are banned from docking within 5 mi of shore, the use of offshore terminals would proliferate, and accordingly, they would become economic because of their regional monopolies on the importing of oil, MMS said. The ports are likely to become monopolies because it would become "exceedingly costly in terms of actual capital outlays and in terms of environmental degradation—due to duplicate pipelines—to have more than one offshore port serving the same refineries in any one area," MMS said.

Other proposals for similar deepwater terminals off Texas never got anywhere. A nationwide mandate for deepwater terminals would cost billions of dollars and invite a whole new round of environmental litigation and regulatory delays.

Double Hulls

• •

The most advanced proposal in the area of tanker safety is actually an old idea that received new life after the Exxon Valdez spill and again after the American Trader spill: requiring double hulls and/or double bottoms on all new tankers.

Fitting a tanker with an extra hull or bottom would, say proponents, help to prevent cargo tanks from being punctured in low-energy tanker accidents, thus reducing the risk of spillage. They point to a recent Coast Guard staff report citing the effectiveness of double hulls in groundings involving LNG carriers—where double hulls are required.

However, the industry generally has opposed double hulling in the past. Opponents raise the concern that double bottoms might make it harder to salvage a grounded vessel. Once the space between the double bottoms of a grounded tanker floods, the theory goes, the vessel loses buoyancy and could be forced deeper into the water. That would cause the vessel to settle more firmly on a reef or other surface on which it ran aground. It would then make it harder to salvage the vessel or keep it stable, thus possibly causing more oil to spill.

Another problem opponents of double hulls see is the possibility of oil vapors seeping into the empty space between the hulls and creating a high risk of explosion. Proponents contend that could be overcome by installation of inert gas systems to reduce the risk of explosion.

There is one area not in dispute: Double hulls would increase the cost of new tankers. By how much is a matter for dispute. Estimates vary from an added 5% to as much as 25% more. However, a vessel of a certain tonnage with a double hull would have considerably less space for cargo tanks. That means less oil to ship at a time or a bigger vessel to accommodate the same size cargo. Either way, it means more shipping costs in addition to the increased construction costs.

What may well settle the issue aside from mandates by Congress in comprehensive oil spill legislation likely to be drafted in 1990 is a study by the NAS for the Secretary of Transportation on tanker design. The final NAS report was scheduled to be released in November 1990.

Further, double bottoms may be of limited use in an incident like the Exxon Valdez spill. According to Coast Guard commandant Adm. Paul Yost, the Exxon Valdez hit Bligh Reef so hard that a double bottom would not have prevented a spill. However, Yost noted the spill might have been half the size it was if it had a double bottom. Double sides would be a better investment, Yost contends, because most tanker accidents are not groundings but collisions with something other than the bottom.

There is another reason for not rushing into a mandate for double hulls/bottoms. In the view of API Vice-President William O'Keefe, such a move would freeze tanker design technology at current levels—"a fatal flaw." O'Keefe cites a promising new technology developed in Sweden that would use atmospheric pressure to prevent oil outflow when the hull of a tanker is punctured.

The mandate would not necessarily be limited to new tanker construction, either. One of the bills before Congress would call for retrofitting all tankers and tank barges that service American ports with double hulls and bottoms. In all, that is almost 2,300 vessels. The track record for retrofitting is hardly extensive: It's only been accomplished once before.

API put the total cost of retrofitting the vessels that call on U.S. ports at $16.6 billion. It also contends that there would not be enough space in U.S. shipyards to meet a congressional deadline that all new ships have double hulls by 1997.

The consequences of rebuilding as much as one-third of the world's tankers and most U.S. barges within 7–15 years would have widespread and unprecedented consequences, according to Brent Dibner, a vice-president of the Lexington, Massachusetts, international management consulting firm Temple, Barker & Sloane.

The resultant surge in demand for shipbuilding at a time when the world's shipyards already have a very high utilization rate will have major financial implications for shipbuilders, shipowners, and oil companies, Dibner said.

According to Dibner, it would be difficult to justify the costs of reconstructing the U.S. tanker fleet, especially with dwindling Alaskan production. That could mean even greater reliance on foreign shippers and oil supply sources, he contends.

Some in the industry maintain the double hull mandate and other oil spill measures being considered in Congress could effectively drive the oil industry's major players from the shipping business. That could create a boom in registering vessels under the flag of remote, unconcerned, or even unfriendly nations. Suppose a spill occurs, and the tanker owner or shipper is out of U.S. jurisdiction? One of the key elements in new oil spill legislation is liability. That is difficult enough to ascertain in U.S. courts. What if the person or company deemed liable in a spill cannot even be found, much less brought to justice in a U.S. court?

Boycotts of U.S. ports by tanker owners in 1990 began shortly before the Mega Borg accident, getting under way in earnest in the weeks that followed the June 8 explosion and fire that ripped through the Norwegian tanker 57 mi off Galveston, killing four and injuring 17 crewmen. Although there was no environmental damage from the resulting spilled oil—most of which burned, evaporated, or sank—the incident shot the issue of tanker safety to the forefront of industry concerns again. About the same time, the British BT Nautilus tanker spilled almost 6,200 bbl of No. 6 fuel oil into New York Harbor, shortly after Exxon announced its new safety program there.

Also at the same time, Royal Dutch/Shell Group banned its fleet of tankers from calling at U.S. ports, limiting calls to LOOP. Royal Dutch/Shell said it was implementing the ban because uninsurable liability in the event of a spill in the United States resulted in risks from the trade that far outweighed potential reward. That was followed by other tanker owners representing fleets totaling more than 100 vessels as this book went to press.

Could we see a situation where oil supplies in this country get squeezed because too many tankers are boycotting U.S. ports over the liability issue? Richard Golob, publisher of *Golob's Oil Pollution Bulletin*, does not think there is reason for alarm.

"Looking at the worldwide situation with tanker shipping, it is clear that the United States is a major market that just cannot be ignored," Golob said. "There may be a certain amount of grandstanding in the boycott to influence legislation."

In any event, says Golob, there is likely to be an increase in lightering in international waters off the United States.

Bans and boycotts may be moot issues anyway. The trend is inevitably toward double hulls, according to Golob.

"Many tankers are going to be forced by their insurance companies to change their designs. Further, with Det Norsk Veritas and Lloyd's Register (accreditation concerns) specifying double hulls as the standard for environmental acceptability, that will affect the major oil companies that are still transporting oil into the United States. They will not allow their cargo to be carried in substandard vessels," he said.

Again, the political heat from the Exxon Valdez and tanker accidents since then is leading politicians and regulators at the federal and state levels to push for a quick fix solution. Irrespective of the wisdom of such a move from an economic or scientific standpoint, the momentum seems inevitable.

At least one company is not going to wait for Congress to act. As part of its series of environmental initiatives announced early in 1990, Conoco Inc. plans to have double hulls and double bottoms installed on all new tankers it orders. That sort of action might help defuse the drive for a short-term mandatory deadline for double hulls in tandem with a retrofit requirement. If the NAS study does not establish a clear and conclusive preference for or against double skins for tankers, an initiative like Conoco's may prove industry's only option in heading off legislative or regulatory action that could prove far worse than double skins for new tanker orders. Again, another spill like the Exxon Valdez or American Trader could kill any choice U.S. oil companies may have in the matter.

At presstime, House and Senate conferees working on U.S. oil spill legislation had agreed to require most oil tankers calling on U.S. ports to be fitted with double hulls by 2010. In essence, the compromise bill mandates that all new tankers be built with double hulls and existing tankers be retrofitted or retired during the next 20 years. Since the conferees melded a stricter House version with a bill supported by the Bush administration, the resulting compromise on the double hull issue would likely be blessed by Congress and the President should an oil spill measure pass in 1990.

The new oil spill legislation will:

- Require double hulls for existing tankers for the period beginning January 1, 1995, at first covering 28-year-old, single-hull tankers larger than 30,000 gross tons and 40-year-old vessels smaller than 30,000 gross tons.
- Allow tankers with double bottoms or sides but lacking full double hulls to stay in service five years longer than single-hull tankers of the same size and age.
- Not require tankers that lighter oil to smaller tankers more than 60 miles offshore or inland water barges to be double-hulled.

Probably the only thing that will blunt the drive toward double-hulling is an accident involving a double-hulled tanker that proves the industry's point: that a double hull could aggravate environmental or safety risks further in certain circumstances. That not only would add regrettably to the transportation industry's poor recent track record, it would be unlikely, according to the Tanker Advisory Center, New York.

About half the tanker spills in 1988–89 could have been prevented or minimized if the vessels had been double hulled, TAC said. Of the 43 spills from tankers of 10,000 DWT or more in 1988–89, double hulls would have prevented or minimized the effects of 28, according to TAC. The rest of the oil spilled involved sinking of loaded tankers after an explosion, fire, or in bad weather. At the same time, TAC acknowledged, double-hull tanker construction costs would be 25% more than those for single-hulled tankers and would raise shipping costs by about 11%.

Lloyd's takes a different view of the situation, concentrating on personnel issues. One Lloyd's official was heard to say that the industry should be concentrating on "double captains, not double hulls."

At the Institute of Marine Engineers conference in London in May 1990, officials from Lloyd's and Shell said that response of watch officers to a developing crisis is more important than ship design.

They maintain that even the very best design and construction would not prevent pollution in a severe grounding or collision. The problem is best dealt with through proper training of the bridge team and implementing a back-up decision-making process in confined waters, wherein no individual could make a decision jeopardizing the ship without confirmation from another officer.

Still, humans will continue to err, so it is wise that industry do everything it can to be prepared for the next big spill.

PIRO and Contingency Plans

PIRO represents the U.S. petroleum industry's best safeguard that the next big oil tanker spill in U.S. waters will have an outcome like the American Trader spill instead of like the Exxon Valdez spill: a quick and effective cleanup leaving behind good will versus the nightmare off Alaska.

Under PIRO, enough equipment will be stationed at major U.S. ports to control a 200,000-bbl spill within five hours in any of five regions (Fig. 9–2), or the equipment could be moved en masse to combat an even larger spill in one region.

The API task force that drew up PIRO also pledged to spend an additional $30–$35 million in the next five years on tanker spill research and development.

The task force also called for extensive reexamination and updating of tanker construction, manning, and operating procedures.

PIRO itself will cost $70–$100 million to establish and $30–$35 million each year to operate. It is to be funded by a fee of 2¢–3¢/bbl on tanker shipments. PIRO members expect the program's initial capability to be in place by 1991 and the full program to be on stream the following year.

Under the program, PIRO will place itself and its equipment under control of the Coast Guard. That will eliminate the conflicts and confusion over lines of authority that plagued the Exxon Valdez response.

PIRO will be implemented by PIRO Implementation Inc. (PII), the organization charged with establishing the program. According to PII Pres. and retired Coast Guard Vice Adm. John D. Costello, PIRO's charge is three-fold:

"First, to build a decisive, robust response organization capable of handling catastrophic spills. Second, to administer a comprehensive research and development program to improve the technology used to respond to spills. And third, to develop industry standards for oil spill response equipment, techniques, and training," Costello said. "PIRO's goal is to have the right people and the right equipment at the site of a spill at the earliest time practicable re-

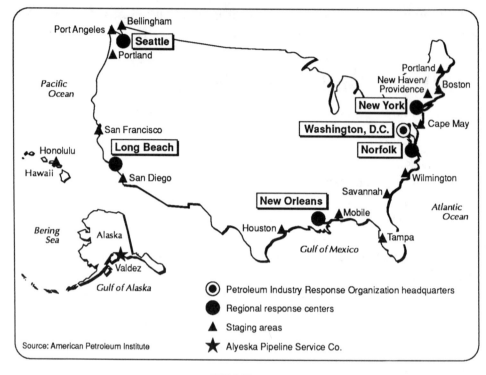

PIRO Elements

Regional center oil spill response capability

Handling a 200,000 bbl spill
Equipment totaling about $15 million
Two 65 ft vessels
16 skimmers with total capacity of 7,400 bbl/hr
64,000 ft of containment boom
10,000 ft of protective boom
22,000 gal. of dispersants
4 pumps
4 fenders
5 dracones for lightering

Total PIRO oil spill response capability: 1.8 million barrels

Figure 9–2 API's proposed United States oil spill response sites.

gardless of time of day, weather conditions, or the remoteness of the scene."

PIRO is not intended to supplant local and regional spill response and cleanup capabilities, but to supplement them in the case of catastrophic spills—leaving the small spills to existing cooperatives and contractors.

Paralleling PIRO is Alyeska's own beefed-up tanker oil spill contingency plan.

Alyeska developed its $45 million/yr oil tanker spill prevention and response plan within a few months of the Exxon Valdez spill. Dubbed the Ships Escort/ Response Vessel System, it focuses on outbound tankers being accompanied by an escort response vessel (ERV) and a ship-assist tug from the Valdez terminal to Prince William Sound's Hinchinbrook Entrance. The program also calls for an 8-vessel, 85-person ship escort and oil spill response organization to be on call at all times.

Alyeska's spill response guidelines now call for it to:

- Contain the oil near its source, recover it mechanically or burn it on the water, lighter remaining oil from the tanker, and keep the oil away from shorelines and open water.
- Treat any oil that escapes from the spill site with dispersants, burning, and mechanical recovery. Keep the oil away from shores.
- Place protective booms on nearshore areas, use mechanical recovery techniques, and clean any contaminated shoreline.

Alyeska also has taken an unprecedented step—for an American company— in accommodating environmental concerns: It has agreed to give extensive influence over its operations to a citizens' watchdog panel. The company will spend $2 million/yr to fund a permanent 15-member group with the power to monitor and investigate its shipping operations. Under the agreement, Alyeska has no say in the appointment of members, cannot reduce the group's funding, and must submit any regulatory or technical disagreements with the watchdog group to binding arbitration. This approach in turn is modeled on a similar program in the United Kingdom North Sea, and may well prove to be the industry's best bet for heading off damaging new regulations and legislation on preventing and containing oil spills.

Exxon's New York Harbor tanker safety initiatives were developed after the company shut down its marine operations there during spring 1990 while it reviewed operations following a string of four spills in the first three months of 1990.

Exxon will spend more than $10 million the next three years to improve environmental and safety performance at its Bayway refinery at Linden, New Jersey, and Bayonne, New Jersey, terminals.

Exxon conducted a study that made recommendations designed to improve vessel maneuvering safety and ensure that only the highest quality vessels call at the refinery and terminal. It plans measures to improve reliability of cargo transfers and boost response capability in the event of an accident. Exxon also created a technical support group for all harbor marine operations.

Exxon will:

- Require tug escort into the harbor for larger vessels.
- Add another tug to the two assisting loaded tankers that move from anchorage to an Exxon terminal.
- Develop tighter requirements and evaluations of third-party tankers.
- Create standards for third-party tugs and barges and monitor how well these craft live up to the standards.
- Install high-level alarms to prevent overfilling and toughen standards for testing and replacing cargo transfer hoses.
- Install more emergency shutdown systems and emergency stop buttons.
- Improve firefighting capability with an elevated foam system at Bayway and by stationing an Exxon tug equipped with firefighting and spill boom deployment capability in the harbor.
- Install two new berth fendering systems.

New Legislation

There is little doubt that, after stumbling around the issue of tanker spill liability since the mid-1970s, Congress in 1990 will have produced some sort of tanker oil spill legislation.

Whether a bill emerges that will survive a Presidential veto remains to be seen. The major stumbling blocks are the question of double hulls/bottoms and the issue of liability.

What is a virtual certainty is that there will be tighter controls on tanker operations.

Likely elements of a comprehensive act establishing policy on tanker oil spill prevention, response, cleanup, and liability are shown in Table 9–1.

It is unlikely the final version of a federal oil spill bill will preempt state liability laws. Some coastal states have adopted standards placing unlimited liability on oil spillers. The congressional leadership likes the deterrent nature

Table 9–1 Likely Elements of 1990 Tanker Safety/Oil Spill Legislation

$1 billion oil spill fund built by a 5¢/bbl fee on waterborne oil.

Oil tankers of more than 3,000 gross tons liable for the first $20 million in damages; total liability limited to $1,200 per gross ton of the vessel.

Tankers and barges of less than 3,000 tons liable for the first $2 million; total liability limited to $1,200 ton.

Liability split equally between vessel operator and cargo owner. Liability not limited on spills caused by gross negligence or willful misconduct.

A phased period for requiring double hulls on new tanker and barge orders (described in text).

Local spill contingency plans for all coastal areas, with compliance by vessels and facilities operating in those areas.

Require tug escorts for tankers in especially sensitive areas.

Require the Coast Guard to issue regulations establishing minimum standards for hull thickness of tankers and alarms to detect leaks and overfills. The Coast Guard also must propose which U.S. ports need new or expanded vessel traffic service systems.

Deny licenses for tanker personnel with drug or alcohol problems.

Require a minimum number of persons on a vessel bridge when in pilotage waters unless a state pilot is on board. In sensitive or hazardous areas, a pilot is required.

More federal funding for research of oil spill response and cleanup technology.

Regional cleanup teams and a database of equipment and contractors around the United States.

Require proof of insurance or financial responsibility to cover maximum liability.

---

of that idea, but a growing tanker boycott could prove to be more than political grandstanding if unlimited liability becomes the federal standard. The question is whether that standard applies in the case of simple negligence or gross negligence. President Bush is likely to veto any bill offering up unlimited liability in cases of simple negligence. So it is likely that a final bill will contain caps on liability that may be onerous but still survivable.

There is likely to be a continuing push to tighten standards and qualifications for personnel directly or indirectly involved in tanker operations, as well as at terminals and storage facilities. The U.S. Coast Guard has recommended increased use of local pilots in the Puget Sound and Columbia River areas of Washington and Oregon. And because the Coast Guard concluded the American Trader spill off Huntington Beach, California, resulted from a lack of knowledge of the correct water depth at the sea berth on the part of the terminal operator and the mooring master and/or his employer, it will pursue new guidelines to ensure minimum keel clearance and tighter mooring master pilotage requirements. The Coast Guard investigation found that the vessel's draft was 43 ft in 4–6 ft sea swells in a sea berth that was thought to be at least 56 ft deep. It proved less than 50 ft deep, causing the American Trader to hole itself twice with its anchor.

Whatever Washington's response to the issue, it is clear that the U.S. petroleum industry faces a radical change in the way it conducts the business of transporting petroleum resources in U.S. waters.

There may be some salvation on the horizon in the implementation of a new technology that shows great promise for mitigating tanker spills: bioremediation.

Bioremediation's Promise

Perhaps the most exciting development in oil spill cleanup technology of late has been the impressive showing of bioremediation in cleaning up oil spills that have reached shore and at sea.

Test results from lab tests and from applications in cleaning up the Exxon Valdez and Mega Borg spills suggest that the oil transportation industry may have a tool for safely and effectively cleaning up oil spills before they can do severe environmental damage.

In the Mega Borg test, the first such test at sea, researchers used a mix of microbes and fertilizer that significantly reduced oil concentrations in water samples collected after the tanker spill. A microbial agent developed by Alpha Environmental of Austin was mixed with fertilizer and seawater in a 55-gal. drum and then sprayed on the surface of the slick in a slurry of about 3 lbs/ac. Treatment took about 30 minutes. Although they didn't collect quantitative data in this test, researchers noted treated oil changed from a continuous brown film and sheen to discrete areas of mottled brown and yellow material. A second controlled test showed consistently that microbial/fertilizer-treated samples had lesser oil concentrations than the control samples.

The Port Aransas test by the University of Texas Marine Science Institute showed microbes reduced oil concentration by as much as 99.99% in a column of seawater.

Exxon, about midway through bioremediation of about 400 shoreline sites in Prince William Sound, said that data from three sites monitored for six weeks by many agencies and companies showed that naturally occurring microbes consumed oil three times faster when treated with liquid and granular fertilizers, all without harm to wildlife or the ecosystem. Exxon lab tests showed that nutrients and oxygen penetrated as deep as 2 ft in some high-energy shorelines, enough to allow microbes to degrade subsurface oil.

Perhaps a wise course would be heavy industry funding for research into bioremediation of offshore oil spills. Perhaps industry could develop a microbial bioremediation capability that could even be deployed from tankers or escort tugs immediately upon a spill's occurrence. Perhaps a bioremediation delivery system could be installed on the tanker in such a way as to allow automatic release with detection of a skin rupture. Would it not be a boon for the tankering industry if it could assure the public it has a quick, clean, safe, and effective method of cleaning up spills? Would that not head off more costly governmental solutions that mandate design and operation practices?

While they are at it, a review of safety procedures and design safety might well be in order for the LNG carriers, which will be in increasing demand in the 1990s.

Pipeline Issues

• •

U.S. pipelines face a similar if less extreme change in the way they operate in regard to spills or leaks. There are not as many variables involved in pipeline transport, but pipeliners and operators of storage and transshipment facilities could easily have their own Exxon Valdez if some key measures are not taken.

Although an onshore oil or products spill generally is not viewed as the kind of major environmental threat that a marine spill is, there remains the potential risk of a catastrophic accident causing widespread environmental damage or even grim repercussions in human terms.

Alyeska has embarked on the most expensive corrosion repair program in industry's history with a campaign to inspect for corrosion, repair it, and replace pipe sections on TAPS.

Cost over the next five years is estimated at $600–$800 million. External corrosion on TAPS and internal corrosion at some of the 800-mile line's pump stations was discovered with the help of a new corrosion detection pig that went beyond the standards for corrosion detection and mitigation that federal law requires. Alyeska also has a major storage tank inspection program under way at Valdez terminal.

Why the extra effort? In the wake of the Exxon Valdez, a large spill in a remote, pristine area of Alaska—even if the spill could be easily contained or cleaned up—would probably draw the same kind of public and political reaction that the tanker spill did.

That also applies to operators of other pipelines, whether liquids products, petrochemicals, or natural gas.

U.S. companies should not take an "It can't happen here" attitude regarding an incident like the horrific June 1989 pipeline disaster in the Soviet Union. In that incident, a natural gas liquids pipeline ruptured, creating a dense vapor cloud that exploded near the Trans-Siberian Railroad. The blast killed more than 600 persons on two passing trains.

Imagine if a leak of a highly volatile product caused a similar tragedy in a heavily populated urban area in the United States. There would be the same kind of heavy-handed rush to legislate quick-fix solutions for pipelines, requiring extensive monitoring and supervisory controls, costly and redundant safety equipment, and punitive liability measures such as those now sweeping the tankering business.

A harbinger of that kind of situation is the controversy involving what measures are needed to keep fishing vessels from striking pipelines on the seabed. That stems from an October 1989 accident in which a fishing boat struck a gas pipeline off Port Arthur, Texas, resulting in 11 deaths, destruction of the ship, and heavy damage to the gas line. The National Transportation Safety Board is investigating and its report likely will contain recommendations that burial depths and methods for all Gulf of Mexico pipelines be reviewed as a safety measure for fishing vessels.

Pipelines must share the tanker industry's mandate: Undertake exhaustive contingency plans for possible spills or leaks resulting from flaws or accidents and develop an effective, worst-case scenario plan for responding to such an incident. It also would be wise to incorporate the public into developing and reviewing such contingency plans.

Another smart move would be more funds for research into super-intelligent pipeline pigs like the one that detected the Alyeska flaws.

Again, zero risk is an impossibility, but it also is a goal industry should strive to approach. That is the direction government is moving, whether through deterrent of huge liabilities and criminal penalties or an onerous regulatory regime.

Industry must take the initiative now, or some pipeline's name could join the ranks of Three Mile Island, Love Canal, Chernobyl, and Exxon Valdez.

Other Transportation Issues

Concern over spills and leaks has tended to overshadow another area of concern to the petroleum transportation industry: its air emissions.

Reauthorization of the CAA will bring some key changes for operators of pipelines, tankers, and terminals.

If a gas pipeline compressor emits more than 250 tons/yr of nitrogen oxides, it is considered to be a major source under EPA guidelines.

Currently, NOx levels allowable from gas-fired turbine compressors are at about 150 parts per million (ppm). EPA is likely to pursue a much more stringent standard, akin to those in California. Its Region 6—the south central states—seeks a 25 ppm standard.

Pipelines with compressors or heating units will face increased monitoring for fugitive emissions of NOx and volatile organic compounds from joints or valves. Efforts are already under way in California—soon to be followed, no doubt, by EPA—to require costly selective catalytic reduction controls on compressor turbine engines.

On the tankering side, the Coast Guard is proposing measures to combat the problem of inert gases and hydrocarbon gases being vented to the atmosphere. The problem occurs when a tanker uses scrubbed carbon dioxide from its boilers to fill empty oil tanks to prevent buildup of volatile hydrocarbon gases. As they fill with oil during loading, the tanks vent the CO_2. At the same time, hydrocarbon gases are emitted from the oil being loaded.

What the Coast Guard proposes is to require construction of onshore vapor emission control systems that would move inert gases by pipeline to a processing plant. Each such plant would cost about $7 million, and each tanker conversion to accommodate such systems would cost about $1 million.

According to John Dunn, managing director of Papachristidis Ship Management Services in London, such a system trades off a minor environmental gain for a major increase in the risk of a tanker accident. Dunn maintains that the pipeline would have to remove the gas from a big tanker at the rate of 25,000 cu m/h. A problem at the recovery plant could result in an instantaneous pressure pulse racing back to the vessel that could tear it apart, resulting in a major spill, according to Dunn.

A taste of what might be in store for U.S. pipelines on air quality issues is what happened to Alyeska in a run-in with Alaskan state regulatory authorities.

Alaska's Department of Environmental Conservation (ADEC), based on a two-year review, concluded that TAPS and the Valdez terminal emitted considerably more air pollution than expected.

Backed by an EPA notice of violation, ADEC claimed the company increased pollution levels by modifying fuel consumption at pump station gas turbines, increased release of hydrocarbon emissions through venting from storage tanks, and increased sulfur dioxide levels by expanding capacity of waste gas incincerators.

The point is not the level of pollution; the point is that ADEC maintains Alyeska should have sought a prevention of significant deterioration (PSD) permit with each of the pollution increases. PSDs require an operator to conduct air quality analyses and apply best available control technology (BACT). Without going into detail, suffice it to say that Alyeska contends there was no pollution increase that warranted a PSD notification and that the claims amount to a regulatory technicality.

At presstime, the agency and Alyeska were still trying to hammer out an accord to avoid a sizable cut in TAPS throughput to accommodate recommended air quality standards.

If successful, ADEC's push for BACT standards could spread to Lower 48 pipelines. That will prove extremely costly.

That is in addition to ADEC's proposed regulations governing hydrocarbon emissions during tanker loading operations at Valdez, which alone will cost the industry more than $100 million, according to Alyeska.

Another "sleeping giant" of an environmental issue is that of contamination of pipelines by polychlorinated biphenyls (PCBs). The issue got widely publicized with the 1987 controversy over PCB contamination at discharge pits along the rights-of-ways of Texas Eastern Transmission Corp.'s gas pipeline system. Texas Eastern agreed to pay a civil penalty of $15 million and spend as much as $400 million over 10 years to clean up PCB contamination at compressor station sites on its system.

The industry can expect more efforts by the EPA to sample and monitor for PCB levels along the nation's gas and liquids pipelines. Companies should proceed toward monitoring and mitigation efforts now to avoid the kind of penalties Texas Eastern has incurred.

Even the wetlands issue and wildlife protection concerns come up with pipeline construction. Getting permits for construction in certain sensitive areas will become increasingly difficult, even for gas pipelines being proposed to satisfy growing demand for gas stemming from air quality concerns.

The Champlain gas pipeline project to the U.S. Northeast—where utility demand for gas is getting desperate because nuclear power is stymied and coal burning contributes to acid rain—was nevertheless hampered by environmental concerns in New England.

The upshot is that the petroleum industry faces its greatest urgency in the transportation sector when it comes to environmental concerns. That is the legacy of the Exxon Valdez. Global warming, acid rain, and groundwater contamination tend to be invisible issues and obscure, long-term threats. But footage of a cute sea otter covered with crude and in its death throes tops the evening news and gives that environmental issue a tangible, gut-wrenching immediacy. That is what made the Exxon Valdez the catalyst it has become. That is why there is no more urgent mission for the petroleum industry than to recapture the public trust in its ability to move petroleum resources safely and with minimal effect on the environment.

10

STRATEGIES FOR REFINER-MARKETERS

What fuels will power American vehicles in the 1990s? A few years ago, the question would not have been asked. The presumption had always been that conventional gasoline and diesel would continue to dominate the motor fuels market well into the 21st century. At some point in the middle of the next century, popular wisdom had it, advances in new energy technologies and vehicle designs would allow the gradual transition from conventional motor fuels based on increasingly scarce reserves of low-cost oil.

That is no longer a safe assumption, because of rising concerns over air quality. The summer of 1988, which brought record ozone levels in some major urban areas along with drought and heat, made the reauthorization of the CAA a pressing item on the nation's agenda. Concerns over vehicular emissions' contributions to global warming and the Exxon Valdez spill gave environmental groups a perfect rostrum from which to target American motorists' unslakeable thirst for gasoline as the major source of many of our environmental ills.

That became clear in the early models of CAA legislation that the Bush administration put forth. Much of the driving force for the administration's approach to cutting air pollution seems to have been inspired by Bush's experience in observing California's "clean fuels" program as Vice-President and head of President Reagan's alternate fuels task force. These mandates were softened somewhat in subsequent versions of CAA legislation (Table 10–1).

Just as the California Energy Commission and later the Scaqmd seemed to have settled on methanol as a panacea for solving air quality problems in that state, so did the early form of Bush's clean air program seem to tilt to methanol.

In the debate that ensued, competing alternate fuels came to the fray to

Table 10–1 Proposed Alternate Fuels

Issues	Senate	House
Scope	Nine cities with serious pollution	Pilot program, California only
VOC emission cuts	All new cars, on a per vehicle basis: Phase 1 (1995–98), 0.75 gpm; Phase 2 (1999 on), 0.66 gpm.	Phase 1 (1995–2000): 50% cut in ozone-forming VOCs. Phase 2: (2000 on): 75% cut in VOCs.
Toxics Definition	Five carcinogens	Five carcinogens, but EPA leeway to add other substances.
Reduction	Phase 1: 12% cut in incidence of cancer cases; Phase 2: 27%, EPA can trim to 18%.	Phase 1: 50% cut in emissions of 5 carcinogens; Phase 2: 75% reduction.
Vehicle phase-in	Applies to all new vehicles from model year 1995 on.	Requires sale of 150,000 vehicles/year in model years 1994–96 and 300,000 in model years 1997 on.
Opt in, opt up	California can develop own opt in, opt up program. Other areas can opt up to California standard. Possible expansion to major corridors.	None
Fuel supply	Must distribute fuel for sale to ultimate buyers specified by EPA.	Enough fuel so vehicles can operate exclusively on clean fuels in the area.
Service stations	Stations of 50,000 gal. Same per month or larger must offer at least one clean fuel for sale. No tank replacement required if tanks have been replaced within last seven years.	

prevent methanol from quickly becoming the only game in town: compressed natural gas, propane, ethanol, and electricity.

Regulatory officials in California seem bent on eliminating gasoline and diesel fuel from the California market by the turn of the century as the only solution to air problems. In fact, methanol is seen as the transition to hydrogen/solar/electric vehicles after the year 2000 in Scaqmd's Air Quality Management Plan (AQMP).

Recognizing the potential loss of a vast market for gasoline—and perhaps the first steps towards its national elimination—U.S. companies, led by ARCO, have responded to the challenge in a way that should be a paradigm for the industry on all environmental issues in the 1990s.

ARCO and other refiner-marketers have ensured that gasoline and diesel will continue to be considered in the equation that is the nation's fuel mix by changing the composition of their fuels. Anti-oil forces cannot argue against fuel choices on grounds of air quality when reformulated conventional fuels' emissions come within the limits established to pave the way for alternate fuels. The question then comes down to a matter of cost and convenience, and there conventional fuels are the clean winner.

Still, widespread refining and marketing of reformulated fuels will not be a simple, inexpensive solution. Refiner-marketers will have to contend with rapidly changing specifications to accommodate national and state concerns over vehicular emissions. And revamping refineries to produce those new products will be costly and difficult to permit. Refineries' own emissions will be a source of consternation in that regard.

At the same time, refiners must contend with a slate of crudes that is increasingly heavier in gravity and higher in metals content. In addition, U.S. motorists' demand for higher octane, premium fuels does not appear to be abating.

U.S. refiner-marketers also must face a host of other environmental concerns that will provide bigger growth in capital outlays than any other aspect of business (Table 10–2). In addition to direct outlays for environmental mitigation, refiners will see an increasing cost component as domestic producers' own environmental costs rise. As the U.S. production slide continues, aggravated by mounting environmental costs not shared by foreign producers and refiners, it will lead to a similar slide in U.S. refiners' ability to compete.

Overall, demand for petroleum products in the United States is expected to continue to climb, albeit at a more moderate pace than in the past few years. It is unlikely that demand will be of a sufficient level to warrant building new grassroots refining capacity to any significant degree in the United States, however. Given the permitting constraints and high costs of building a grassroots refinery in the United States today, the availability of low-cost imports probably will keep that from happening.

It is more likely that U.S. refiners will be able to meet expected U.S. demand growth through debottlenecking and incremental capacity expansion ("capacity creep").

However, that potential demand growth could be stifled early in the game if separate efforts to search for budget deficit solutions and to tax fossil fuels on the basis of their air emissions eventually converge. Given the "motherhood and apple pie" grassroots appeal of the environment these days, perhaps a

Table 10–2 U.S. Refiners' Projected Environmental Spending in the 1990s

	Million dollars/yr
Near term	
Underground tanks	210
Landfarming refining wastes	106–610
Listing wastewater treating sludge as hazardous	2,000
Right-to-know reporting	55–75
Gasoline volatility limits	1,500
PSD for NOx	1–100
RCRA action for solid waste disposal	300–800
Cercla outlays in excess of Superfund tax	9,000
Subtotal	**13,173–14,285**
Probable near term	
Diesel fuel 0.05% sulfur limit	1,200
Gasoline minimum oxygen requirements	15–200
New benzene emission limits	Not estimated
RACT for VOC, NOx	3,970
New refinery effluent guidelines	Not estimated
New transportation related requirements	Not estimated
Management of used oil under RCRA	Not estimated
Subtotal	**5,185–5,370**
Possible long term	
Acid rain/SOx[2]	100–2,000
Toxic air pollutants	45–200
Offshore air emissions	170
Added Stage II vapor controls	200
Methanol fuels	Not estimated
Aboveground tanks	Not estimated
Waste minimization	Not estimated
Offshore effluent guidelines	Not estimated
UIC rules on production wastes	Not estimated
RCRA rules on solid production wastes	Not estimated
Global warming/greenhouse effect legislation	Not estimated
Subtotal	**550–2,590**
Total excluding items not estimated	**18,873–22,245**

Source: Crown Central Petroleum Corp., in testimony before the Department of Energy.

weak-kneed Congress and even President Bush could sugarcoat an even bigger gasoline tax hike enough to swallow it.

As capacity tightens in the 1990s and demand is sustained at a fairly healthy level, the general outlook for U.S. refiner-marketers is one for good margins and profits, even with the soaring outlays for environmental mitigation.

However, if some sense of balance is not maintained in formulating U.S. energy and environmental policies, then U.S. refiners will see their competitive position eroded to the point that they can no longer compete with foreign refiners. Hence, more refining capacity will move abroad, and the United States will supplant its dangerous dependence on foreign crude sources with an even more dangerous dependence on foreign petroleum products sources.

Competing Fuels

· ·

Much of what will be on the clean air agenda for refiner-marketers was covered in Chapter Four.

However, it is instructive to take a closer look at the competing alternate fuels and the flurry of studies that tout the benefits of one over the other.

U.S. refiner-marketers will continue to hold their own with conventional fuels to the degree that they are able to keep the clean air/clean fuels debate focused on air standards and not mandated fuel choices. It is simple enough to mandate a fuel choice by setting an emissions standard that automatically excludes conventional petroleum fuels—as the Daschle CAA amendment does for ethanol. It is then the mission of refiner-marketers to point out to the public the uncertain health benefits—and perhaps even hazards—and the staggering economic consequences that the public faces under new fuels mandates.

According to innumerable polls on the subject, U.S. motorists are willing to pay a price for cleaner air. The question is, how much? Would they accept a price tag of another 50¢/gal. for cleaner fuels that might have been required under California's "Big Green" initiative? What if it doubled the price of motor fuels, or even trebled it? An indicator of that could be seen in the runup of prices in summer 1989, when temporary shortfalls of crude owing to the Exxon Valdez spill and supply disruptions in the North Sea combined with higher costs for environmental mitigation led to price spikes of as much as 25¢/gal. in a few months. Motorists' overall demand was affected only slightly, but there was a large-scale switch back to less premium grades of gasoline as price escalated.

Further, what kind of price would U.S. motorists put on health and safety? Methanol, perhaps the leading candidate to replace conventional gasoline and diesel, is a deadly poison that has an extremely high affinity for water, can be rapidly absorbed into the skin, and is a cumulative poison. Only a tablespoon of the stuff can kill.

What if methanol was mandated throughout the country, as it seems to be on course in California, and there followed a rash of poisonings that triggered the same kind of hysteria that almost put the apple industry out of business over Alar? Could the refining industry, as the likely producers and distributors of a methanol-based fuel, survive, especially after the cost of a transition to methanol?

It seems that the welter of legislative and regulatory proposals to find a solution to motor fuels' contribution to atmospheric pollution could become a headlong rush into a public policy disaster along the lines of the synthetic fuels program during the late 1970s and early 1980s.

Methanol Issues

There are a number of good reasons why methanol is the front-runner among alternate motor fuel choices, as seen by the EPA's 1989 report on the economic and environmental effects of methanol as a motor fuel.

The EPA report found that:

• Methanol is a clean-burning fuel that could be produced from ample supplies of natural gas—the lowest cost source.
• It is competitive with gasoline at current world oil prices. Assuming a chemical grade methanol price of 40¢/gal. and making adjustments for methanol's inferior energy content (about half) and superior energy efficiency (about 5% more) versus gasoline, EPA figured that a blend of 85% methanol and 15% gasoline (M-85) would retail for about $1.14–1.24/gal. That would compare favorably with premium unleaded gasoline, considering methanol's high octane rating. Eventually, pure methanol in dedicated vehicles would be the best bet, said the EPA, because of fuel efficiency and environmental benefits. Neat, or 100%, methanol, would retail for 92¢ to $1/gal., the agency estimated.
• M-85 would cut emissions of volatile organic compounds from vehicles by 30% and from neat methanol by 80%. Methanol fuels also would sharply reduce emissions of air toxics, while advanced catalytic converters could trim methanol's emissions of the carcinogen formaldehyde to acceptable levels, EPA said.

The industry does not see the issue the same way. Although a gallon of methanol fuel might compete in dollars and cents with gasoline, it provides only half the miles per gallon. So it follows that the cost of fueling a methanol vehicle might be as much as double that of a gasoline-fueled vehicle.

In addition, the cost of producing methanol on a large scale would be huge. A plant to produce 50,000 B/D of methanol would cost as much as five times that of a comparable 50,000 B/D refinery. That does not include the staggering costs of new distribution, storage, and safety systems.

Vehicles capable of running on methanol fuels would also cost the consumer more. Public opinion polls cited by American Petroleum Institute (API) suggest that only about one-fifth of American car buyers would be willing to spend as much as $1,000 for a car designed to run on a fuel that cuts air pollution.

All of this brings up the question of energy supplies. The argument for methanol from natural gas—methanol from coal would be wildly uneconomic and actually increase relative emissions of CO_2—as a cure for U.S. energy security woes falters when it becomes apparent, as it did in the EPA study, that the Middle East, Australia, and Trinidad would be the lowest cost suppliers of methanol in the 1990s. That means increased tanker-borne shipments of a highly water soluble toxin in U.S. waters and simply displacing U.S. dependence on Middle East crude with U.S. dependence on Middle East methanol. A study by Sierra Research Inc., Sacramento, California, for API indicated that U.S. tanker traffic could quadruple under a broad methanol mandate.

In order to provide enough methanol to supplant equivalent volumes of gasoline in line with U.S. demand growth, methanol manufacturers would have to boost worldwide capacity to 300 billion gal./yr from the current 6 billion gal./yr. Supplying California's total fuel needs alone would call for more than 20 methanol plants at a cost of about $1 billion apiece.

With all that said, would methanol's environmental benefits be worth it? The record is not as clear as the EPA would like to believe. For one thing, tests showing improvement in emissions for methanol versus gasoline involve small-scale laboratory simulations and computer modeling. There is no study showing the actual performance of methanol vehicles in affecting an urban area's air quality. Anything else is largely speculation.

Methanol's chief attraction is its ability to reduce ozone levels. To what degree it is able to do that is subject to dispute, according to different assumptions in the various studies of methanol emissions. Many of those assumptions, according to API, hinge on the level of formaldehyde emissions. Not only is formaldehyde a carcinogen, it has been identified as a major culprit in ozone formation. Thus studies touting the ozone reducing ability of methanol assume that formaldehyde emissions can be stringently controlled over the life of a vehicle. Yet that technology does not exist.

Further, says API, methanol vehicles may emit more nitrogen oxides—an ozone precursor—than gasoline vehicles. In addition, methanol's effectiveness in reducing ozone may vary substantially day to day, depending on the total mix of emissions with a given urban area.

As to methanol's contribution to the global warming debate, a 1988 study by the Congressional Research Service concluded that substituting methanol for gasoline would not change total carbon dioxide emissions if the methanol were made from natural gas.

There is also a drawback in the use of methanol-gasoline blends. A mixture results in a more volatile substance than either neat methanol or gasoline. Because emissions increase with volatility—the reason EPA wants volatility reductions in gasoline—methanol's gain would be lost.

According to a study by Carnegie-Mellon University, if all vehicles in the Los Angeles area ran on neat methanol, by the year 2000, peak ozone levels would drop by 9.2–14.5%—assuming that effective formaldehyde controls existed. By 2010, the decrease in Los Angeles peak ozone under those conditions would be 5.8–11.3%.

What that study overlooks, says API, is that flexible fuel vehicles—running on a mix of methanol and gasoline or switching between the two—would have to be part of the equation because they are required for long distance driving. So the expected emissions reductions may not materialize to the degree envisioned.

Further, the Sierra Research study concluded, the current level of emissions control technology suggests that the gasoline vehicles would have 50% fewer NOx or hydrocarbon emissions compared with methanol vehicles.

There are safety considerations in methanol use as well. Aside from ingestion resulting in blindness or death, methanol can accumulate in the system by absorbtion through the skin. A driver who wore clothing that he had spilled methanol on—say, while fueling his tank—for an extended period could be so exposed.

Although it is harder to ignite than gasoline, methanol presents a different kind of fire safety risk because it burns with an invisible flame in daylight.

Moreover, API concludes, methanol's lower energy content calls for an increase in the capacity of the nation's fuel delivery system by 70–80%. Beyond that, methanol's highly corrosive qualities would force the replacement of valves and service station controls as well as modifications to pipelines and pumping stations.

What is especially disturbing about methanol is that the EPA and others entrusted with safeguarding the public health are willing to risk massive distribution of a substance about which the long-term environmental and public health effects are not known. An internal draft study by the EPA's office of research and development calls for further research into the effects of long-term, low-level exposure to methanol. It cited two studies that provided evidence that exposure of rats to concentrations as low as 200 ppm for as little as six hours resulted in significant reproductive toxicity. That sort of evidence has led EPA to seek bans on certain substances under other circumstances.

It is ironic that the federal government would pursue widespread experimentation of a new fuel that releases significant emissions of formaldehyde, while at the same time the Labor Department is under court order to conduct a new review of the cancer risks of exposure to formaldehyde and to reconsider its decision not to protect the earnings of workers disabled by the chemical.

OSHA in 1987 set a maximum exposure limit of 1 ppm during an eight-hour day. Several unions challenged that as not stringent enough, which sent the matter to court.

There is one area of methanol use that the petroleum industry is wholeheartedly endorsing: the production of the oxygenate methyl tertiary butyl ether MTBE from methanol. MTBE is gaining widespread acceptance as a key component in the U.S. gasoline pool to meet the growing demand for octane and address environmental concerns in its ability to cut carbon monoxide emissions.

Despite the misgivings, the push for methanol still is alive at the federal and state level. Recognizing that, a number of other alternate fuels have been recently jockeying for position to take advantage of any federal largesse that might be forthcoming.

Ethanol and Pork

Among the other alternate fuels clamoring for attention under the clean air spotlight, ethanol has the most lobbying clout despite its economic shortcomings.

Ethanol's chief attribute is its ability to reduce carbon monoxide levels, as seen in tests in Colorado. Proponents also argue in favor of ethanol as a renewable resource, since it can be produced from corn and other crops. So there are benefits for American farmers as well. By providing a large new source of domestic fuel, ethanol also would help U.S. energy security.

If this sounds familiar, that's because it is. The great gasohol campaign of the late 1970s did not result in ethanol being widely embraced by the American consumer, even though it was heavily subsidized by federal and state government tax exemptions. Reports of vehicular performance problems did not help gasohol's reputation, and the upshot is that now gasohol has a significant presence only in the Farm Belt states, where the heavy subsidies remain and parochial interests are served.

Gasohol demand growth has zeroed out since the mid-1980s, and there remains a real concern about ethanol's contribution to the pollution problem in boosting a fuel's volatility. Allowing an EPA waiver on volatility limits and continuing the subsidies on ethanol together is the only way that the fuel will survive at all. The subsidies are critical. It takes more energy to produce ethanol than the energy that is derived from it. From a standpoint of pure economics and energy policy, ethanol would have to be considered a white elephant.

Still, the ethanol lobby persists, trying to have a subsidy extended to an ethanol derivative to compete with MTBE. Ethyl tertiary butyl ether (ETBE) would enable ethanol to retain much of its share of the motor fuels market because it eliminates the volatility problem. Again, there is the question of cost, owing to the feedstock cost. Congressional efforts to have ethanol's 60¢/gal. subsidy extended to ETBE have proven moot. The Internal Revenue Service has already taken that step in tax rulings.

At the same time, the ethanol lobby's point man in Congress, Sen. Tom Daschle (D.-S.D.), was successful in getting the ethanol agenda on the Senate version of the CAA. The Daschle amendment sets standards for reformulated gasoline and requires its use in the nine urban areas that do not meet the EPA standards for ozone. In addition to cutting the aromatic hydrocarbon content of reformulated gasoline to 25% from 33% today and benzene levels to 1%, the amendment would require reformulated gasoline to have an oxygen content of at least 2.7% by weight. It is unlikely that there would be any candidate for achieving that level of oxygen content other than ethanol. In case there is any doubt about the efficacy of this amendment's contribution to the ethanol pork barrel, Daschle estimated that it would more than double the market for ethanol in the United States.

API contends that the Daschle amendment would cost $100 billion and raise the cost of gasoline by 25¢/gal. with a questionable benefit for air quality. Raising oxygen levels in gasoline to 2% by weight helped Colorado's Front Range trim carbon monoxide levels significantly. However, the simple turnover in the vehicle fleet with newer cars with oxygen sensors to maintain the correct air-fuel mix (thereby lowering CO emissions) was a major contributing factor. Industry studies suggest that as this turnover in the nation's vehicle fleet continues, the CO problem will resolve itself by the mid-1990s.

In addition, there is the question of gasohol's other air emissions. Nitrogen oxide emissions can increase by as much as 18% with gasohol use, according to industry studies. Ethanol blended in gasoline also raises the latter's volatility by as much as 1 PSI Reid vapor pressure, which contributes to the ozone problem. By turning ethanol into ETBE, that eliminates the problem of increased volatility. But ETBE then must have a waiver on oxygen content and the federal subsidy.

Refiner-marketers have been selling gasohol in a handful of states for a number of years. It has been the experience of companies like Amoco Corp., which has been among the biggest marketers of gasohol because of its market concentration in the Midwest, that the majority of consumers do not like gasohol and even have intensely negative feelings about the fuel. According to market research by these companies, dislike for gasohol tends to grow the longer it stays in a market.

It would behoove refiner-marketers who do not want to see consumers' choice of motor fuel effectively limited to one they already dislike to trot out the results of that consumer research along with the studies of economic and environmental effects of gasohol use. The average American citizen's eyes may glaze over when confronted with computer models and technical details over photochemical reactions in the atmosphere. But a few well-placed testimonials on television by disgruntled former gasohol users could prove more effective than a host of studies.

At the same time, with Washington's focus on budget deficits and pork barrel/influence peddling scams surrounding the savings and loan debacle, the petroleum industry should not be afraid to get very specific about the costs and benefits of subsidizing the ethanol industry and how little that actually benefits the average American farmer.

Other Fuels

· ·

Gas producers, pipelines, and gas processors have a major stake in ensuring that they are not left behind in the rush to embrace alternate fuels.

While much of the government emphasis and political lobbying has been on alternate liquid fuels, proponents of compressed natural gas and liquefied petroleum gases have been slow to capitalize on their potential in the energy/environment debate.

That is due mainly to the search for a solution that will meet the needs of the average motorist. But there remains a sizable potential for CNG and LPG in the nation's commercial and private fleets of cars and trucks. It took the perceived tilt of the Bush administration toward alcohol fuels to awaken the gas fuels lobbies to more aggressively touting the virtues of their products.

According to the National Association of Fleet Administrators, fleet vehicles account for about 5% of all U.S. vehicles. Converting all such vehicles to LPG or CNG would amount to a big new market. There already is a sizable infrastructure for LPG vehicles in the United States. CNG is making inroads as well, notably among local gas distribution utility fleets and municipal buses.

LPG proponents can point to their product's environmental benefits along with advantages over alcohol fuels. EPA tests of a propane-fueled vehicle showed propane to be as clean as methanol while providing better mileage. The EPA tests showed propane emissions versus gasoline were 93% lower in CO, 39% lower in NOx, and 73% lower in net reactive exhaust hydrocarbons.

LPG associations contend that propane is the alternate clean fuel closest to gasoline in cost per gallon and in miles per gallon. Propane delivers about 85% the mileage of gasoline versus methanol's 54% and ethanol's 70%, based on BTU energy content per gallon, according to the Western Liquid Gas Association.

Perhaps the strongest argument in favor of LPG is its track record, with more than 350,000 vehicles on the road in the United States, mainly in commercial fleets. A clear economic advantage made this infrastructure possible. Gasoline engines can easily be converted to propane or a dual gasoline/LPG use, for only about $750. For vehicles that run about 30,000 mi/yr, that yields a payout in less than three years. LPG fleet operators have noted a sharply reduced maintenance for propane vehicles because the fuel burns more cleanly than gasoline.

Drawbacks to LPG are the weight of the pressurized tank it requires and public perceptions of its safety. A 20-gal. propane tank weighs about 150 lb more than a comparably sized gasoline tank. That may not prove daunting for a large van or truck, but it might for an owner or prospective buyer of a subcompact passenger vehicle.

Consumers tend to get skittish about any fuel that is stored in a pressured tank. The rash of publicity over fires and injuries related to exploding disposable lighters in recent years has not helped mitigate that perception. In fact, propane is no more dangerous to handle than gasoline. And it is less susceptible to explosion or fire in an accident because the thick-walled pressurized tank on an LPG vehicle could withstand an impact better than a thin-walled steel gasoline tank.

LPG advocates must face up to the task of winning consumer confidence in LPG safety and convenience. A nationwide network is in place to manufacture, distribute, and sell LPG. The price is right at about 70¢/gal., considering the high octane and mileage features.

Currently, only about 5–6% of U.S. propane supplies goes toward use as a motor fuel, according to the Gas Processors Association. More than one-third of U.S. propane supplies is used as petrochemical feedstock, mainly because an LPG surplus has kept a lid on propane prices in recent years. Hiking propane use as a motor fuel would place more of a premium on the fuel, raising its price and therefore driving much of the supply out of the petrochemical market and into the motor fuels market.

Surplus gas processing capacity could be put to use as well, doubling U.S. output of propane to 1.7 million B/D. It would not require a massive effort to expand the nation's retail propane marketing network to accommodate non-fleet motorists in the major nonattainment urban areas, as it would for the alcohol fuels.

Infrastructure is a strong argument on behalf of CNG as well. A natural gas pipeline network already extends 1 million mi across the United States. CNG also has a strong track record, with more than 500,000 CNG vehicles in operation worldwide.

Natural gas may be the best alternative as a motor fuel from an environmental standpoint, say advocates of CNG. Converting existing gasoline or diesel vehicles to natural gas can slash emissions of CO by as much as 99%, NOx by as much as 65%, and reactive hydrocarbons by as much as 85%.

According to the American Gas Association (AGA), converting half of the nation's 16 million fleet vehicles could cut U.S. oil consumption by 500,000 B/D, or 5% of total consumption by the transportation sector.

But the natural gas industry has come under some criticism by oil products marketers after AGA stepped up its lobbying efforts in 1989 after passage of the Alternative Motor Fuels Act. The Petroleum Marketers Association of America (PMAA) blasted AGA's efforts to portray natural gas as a "panacea" for U.S. energy and environmental needs. PMAA also claimed that increases in U.S. consumption of natural gas to meet growing motor fuel uses would result in increasing reliance on imports of that fuel, notably in the form of LNG from the Middle East. At the same time, said PMAA, it would take 25% of U.S. gas production to meet 20% of U.S. motor fuels consumption.

That sort of internecine squabbling among different sectors of the petroleum industry isn't necessarily a harmful thing, if it is simply a case of competing forces battling it out for market share on a level playing field. However, when one segment of the industry seeks preferential treatment, it automatically skews the economics of energy use and just invites more government intervention.

There can be room for all competing petroleum fuels, if they are placed on equal footing. Some competitors may even be of benefit to each other, such as the case with diesel and natural gas. Under EPA strictures, NOx emissions in new diesel engines must be halved from current levels by 1991. Bus and heavy truck engines will have to cut particulates emissions to one-sixth of today's levels by 1994.

Correspondingly, the Gas Research Institute is working on a dual fuel conversion kit for diesel engines that involves using diesel as a sort of pilot light, ignited by the heat of compression and in turn firing the natural gas. At idle and low speeds this kind of diesel engine would run on diesel fuel and at higher speeds on natural gas. That would provide some progress toward meeting future emissions reductions goals without eliminating diesel from the market and mandating costly and inconvenient single fuel alternate fuel engine designs.

The Brazilian Model

If U.S. refiner-marketers want to convince American motorists of the folly of government-mandated fuels, they have a perfect model in Brazil's alcohol fuels program.

Brazil launched an $18-billion program in the 1970s to mass produce fuel ethanol from sugar cane. The country at the time was reeling from the effects of a massive oil imports bill due to the price spikes of the 1970s.

Initially, it seemed to make sense: Oil imports would be slashed; struggling plantations would get a boost in a limping sugar cane market; and the economy would benefit from the ripple effects of a new industry. Brazilian law decreed that state company Petrobras must provide gasohol at prices competitive with gasoline, and a certain share of new vehicles must be designed to burn gasohol.

Until 1989, 90% of the new cars built in Brazil ran on gasohol. There were signs gasohol reduced carbon monoxide emissions in the big cities like Sao Paulo and Rio de Janeiro. Brazil's program was touted as a paradigm of an alternate fuels program.

Unfortunately, the alcohol fuels program ran afoul of the oil price collapse. With oil prices halved, the Petrobras subsidy effectively doubled. Petrobras lost $650 million during 1980–89 in its efforts to commercialize mass production of fuel ethanol from sugar cane.

Because of its mounting losses, Petrobras was obliged to slash ethanol prices to $32/bbl from about $60/bbl in 1979. In midsummer 1990, plans called for Petrobras to cut further its subsidy for ethanol and trim its minimum blending requirement—possibly phase out an ethanol mandate altogether.

Petrobras sold its strategic ethanol stocks, and with the subsidies slashed, cane growers have switched back to exporting sugar. Distillers refuse to plant more cane.

Consequently, shortages of fuel ethanol have cropped up in many of the Brazilian states. Many Brazilian drivers were canceling motoring vacations because they feared not being able to find gasohol in many areas. There is a rush under way in big cities like Sao Paulo at auto shops that are hastily converting gasohol-fueled cars back to gasoline.

There are a couple of real ironies here that ought to serve as a centerpiece for the debate with the ethanol/methanol proponents. With the shortfall of ethanol in Brazil, ethanol suppliers have been forced to import methanol to stretch supplies.

Brazil's environmentalists were already unhappy over the destruction of rain forests to allow sugar cane plantation expansions. Now they were aghast over the widespread introduction of methanol into the fuel supply, claiming methanol to be a dangerous, toxic substance.

When a methanol tank truck overturned in a rural area and spilled its cargo near a Brazilian river town, the result was tainted drinking water, a huge fish kill, and sick humans and livestock. The scandal gave the country's environmentalists a lot of new ammunition for outrage.

A U.S. refiner-marketer would be wise to hold this up in the debate over alternate fuels with a cautionary, "It could happen here."

In short, when it comes to the alternate motor fuels industry, whose growth could accelerate exponentially in the 1990s given regulatory or legislative preferences, the petroleum industry would best be served by insisting that government, through legislation or regulation, favor no single fuel over another in meeting reasonable air emission standards.

In an ideal world, the service stations of the 1990s will be "fuel cafeterias," offering a wide range of choices, with the consumer allowed to make the choice on economic and/or environmental grounds. Together with the turnover in the vehicle fleet, there is no reason why significant progress toward the desired clean air standards cannot be made this way.

Reformulated Fuels Push

• •

If the petroleum industry seizes the initiative now, it can ensure that gasoline and diesel fuel remain the "blue plate special" in that fuel cafeteria of the 1990s.

Refiner-marketers already have made significant strides toward that end with the introduction of reformulated gasolines and diesel fuels. Essentially, they are conventional motor fuels stripped of or reduced in the components that yield offending levels of pollutants.

ARCO launched industry's offensive in the battle to head off regulatory campaigns mandating alternate fuels for air quality reasons. In 1989, ARCO began replacing its leaded regular gasoline in Southern California with a reformulated gasoline blend that cuts older vehicles' smog-causing emissions by as much as one-third.

That blend, EC-1, contains no lead, one-third less olefins and aromatics, 50% less benzene, and 80% less sulfur than regular gasoline. ARCO added MTBE to give the blend at least 1% by weight oxygen and reducing its volatility to 8 psi from 9 psi.

The reformulation costs a few cents more to manufacture—which ARCO absorbs—and does not involve significant refining changes. What will prove more complicated are similar reformulations of unleaded and premium unleaded gasoline grades, an effort likely to take three to five years of development and testing. ARCO already has earmarked $2 billion to revamp its Cherry Point, Washington, and Carson, California, refineries to produce reformulated gasolines that can, in tandem with auto design improvements, meet the expected clean air standards.

According to ARCO, the 1.2 million older vehicles without catalytic converters in California's South Coast air basin account for only 15% of the vehicles on the road in the basin but are responsible for more than 30% of mobile source pollution there. ARCO contends that if all older cars in Southern California burned EC-1, the net effect would equate to eliminating 20% of the most polluting vehicles from the road. That would deliver clean air benefits immediately that otherwise would not be felt for several years through turnover of the vehicle fleet.

ARCO projects that EC-1 would significantly reduce ozone in the South Coast air basin by cutting 350 tons/day of reactive organic gases, CO, and NOx. Vehicles using EC-1 emit 4% less hydrocarbons, 10% less CO, 5% less NOx, and 21% less evaporative emissions than vehicles burning conventional leaded regular.

Other refiner-marketers have quickly followed suit (Table 10–3). It is apparent that ecologically correct gasoline will prove to be at least as hot a marketing craze as the campaigns for detergent and high octane gasolines—both of which are likely to continue and even benefit from the push to reformulate cleaner gasolines.

Industry's reformulation campaign also will spread to diesel fuel, out of necessity. EPA has proposed a rule to cut sulfur concentrations in diesel fuel by 80% to curb particulate emissions.

EPA's new rule would require refiners to cut the sulfur content of highway diesel fuel to a maximum of 0.05% by weight from the current average level of 0.25% by weight. Refiners would have to comply with the new rule by October 1993. EPA estimates the rule would add 1.8–2.3¢ to the cost of a gallon of diesel, but the National Petroleum Refiners' Association pegged the tab at several cents per gallon more.

The EPA also proposes to cap diesel aromatics at their current levels by imposing a diesel cetane index specification.

Even if nothing happens with reauthorization of the CAA, refiners face the prospect of reformulating fuels under the regulatory regime of EPA. Pushed

Table 10–3 U.S. Refiners' Reformulated Fuels Programs

Company	Fuel	Specifications	Market
Amoco		Reduced volatility blend (9.5 psi Rvp)	Northern Illinois, including Chicago, parts of NW Indiana
ARCO	EC–1	88 octane, 8 psi Rvp, MTBE/1 wt % oxygen, no lead, 20% aromatics benzene 1%, maximum 300 ppm sulfur, olefins cut by one-third.	Southern California (pre-1975 cars only)
Conoco	RXL regular	89.5 octane, 8.5 psi Rvp-MTBE/up to 2.7 wt % oxygen, 25% aromatics, 2% benzene	Denver, unleaded only
Diamond Shamrock	RG–87	87 octane, no lead, MTBE, 20% aromatics, 8 psi Rvp	Colorado's Front Range
Marathon	Amaraclean	87–92 octane, 10% MTBE/1.8 wt % oxygen, 9.5 psi Rvp, 25% aromatics	Detroit, all grades
Phillips	Super Clean Unleaded Plus	89 octane, MTBE, 33% cut in sulfur, 35% cut in benzene, 20–30% cut in olefins, aromatics	St. Louis
Shell	SU 2000E	Premium unleaded, 5.5% MTBE/1 wt % oxygen, 100 ppm sulfur, 8–8.5 psi Rvp, 40% aromatics, 10% olefins, benzene 1.3% (East), 2.2% (West)	Washington, D.C., and nine worst ozone areas
Sun	Diesel	Reformulated diesel, 0.05 wt % sulfur, cetane rating of 45, reduced particulates	Selected tests in Philadelphia, Baltimore, Florida, New Jersey
Sun	Gasoline	Reformulated gasolines, reduced olefins, aromatics, higher oxygenates levels, reduced volatility	Lab tests at Marcus Hook, Pennsylvania, possible retail sales in second half 1990

into an early effort at mandating lower summertime volatility by a clutch of northeastern states, EPA acceded to the states by not overruling their near-term standards for volatility. EPA's own proposed rulemaking would cut summertime volatility by as much as 3.3 psi by 1992. Costs of this measure range from EPA's estimate of $450 million/yr to the industry's estimate of $1.5 billion/yr.

Seven northeastern states have already jumped the gun on EPA, implementing in summer 1989 a new limit for restricting gasoline volatility to 9 psi. Although there were some fears of potential product shortfalls that year, they did not materialize—anticipated price increases as a result of the rule did, however.

The upshot of still tighter volatility rules could mean increased profitability for U.S. refiners, according to Robert Hermes, president of Purvin & Gertz Inc., Houston. Hermes maintains that stricter volatility rules at a time when U.S. refinery utilization is operating close to maximum seasonal capacity could actually improve margins. By reducing refiners' production—a 1 psi volatility cut results in a seasonal maximum production cut of 110,000 B/D—utilization rates climb and the cost of producing the incremental barrel rises, thus boosting margins, Hermes figures.

States also are leading the way on mandating the use of oxygenates, with mandatory seasonal floors for oxygen content in gasoline in areas of Colorado, Arizona, and New Mexico already in effect. If refiners can convince the federal and state governments to limit oxygen rules to 2% by weight, then they can meet that standard with MTBE. It seems pointless to push for higher oxygen levels when the turnover in the vehicle fleet probably will eliminate the carbon monoxide problem by the mid-1990s. Yet you can look for the ethanol lobby to press for a higher oxygen content in the key target states and even at the federal level. Because EPA limits MTBE use to an oxygen limit of 2% by weight in unleaded gasoline, you can be equally certain that the ethanol advocates will push for a level beyond 2%.

To meet expected demand for octane and required oxygen levels in gasoline, refiners will turn in growing numbers to ethers. According to George Unzelman, president of HyOx Inc., Fallbrook, California, ethers will account for 2.5% by volume of the U.S. gasoline poll by 1995 and perhaps 10% by 2000. MTBE will account for the biggest share of the ethers market—under Colorado's oxygenates mandate, MTBE won 95% of the market—unless continued subsidies shift the economics in favor of ETBE, says Unzelman. Another contender will be tertiary amyl methyl ether (TAME), derived from methanol and isoamylene. If further efforts to cut volatility preclude a waiver for ETBE, then TAME could gain a significant foothold in the U.S. ethers market.

Reformulated Fuels Challenge

. . . .

U.S. refiners will face a continuing prospect of gasoline supply shortages and resultant price jumps in their efforts to produce and market reformulated fuels.

Industry already had a taste of that when some northeastern states jumped the gun on EPA by introducing their own lower volatility limits for summertime gasoline. Regional supply shortages resulting from refiners frenziedly trying to meet octane demand under the new volatility limits contributed sharply to the runup in gasoline prices that was widely blamed on the Exxon Valdez spill.

There are likely to be more instances of that, especially during the summer, in the 1990s. That is because U.S. refining capacity is stretched almost to its effective limits now. There may be some room for debottlenecking to stretch capacity to meet expected demand, but there is not likely to be any relief from construction of grassroots refinery. Such a facility would be hard-pressed to be economic at today's prices and even harder-pressed to win environmental permits.

As was seen with the deadly blast and fire that killed 17 and shut down ARCO Chemical's Channelview styrene monomer/propylene oxide plant, any disruption of supply can affect the market strongly. Because the Channelview plant also had the biggest U.S. facility for producting MTBE, loss of the octane enhancing component shot the price of unleaded gasoline futures by 10¢/gal. the first week or so. That happened even as the market saw an unexpected softness in demand at the start of the 1990 summer driving season.

But what will happen when reformulated fuels become commonplace? And what if the specifications for those fuels are stipulated strictly by government, as is the case with current versions of the CAA (Table 10–4)?

Gasoline demand growth, by most analysts' projections, will be modest in the United States at best in the 1990s. And that demand growth, especially for premium, high octane grades, is acutely sensitive to price increases. Premium grades today offer U.S. refiners their best margins.

Reformulating gasoline to meet air quality goals is certain to boost refiners' costs. But they may be limited to the extent they can pass along those costs. If one region is compelled to use reformulated fuels, the price of gasoline will be set in adjacent regions, and the refiner may have to eat those added costs.

What if the federal government elects to hike gasoline taxes again, now that "Read-my-lips" has proven so ephemeral? The combination of higher gasoline costs passed onto the consumer and an added tax bite could squelch demand for gasoline considerably. At the same time, declining production of

Table 10–4 Reformulated Gasoline Specs under Clean Air Act Legislation

Issue	Senate	House
Refiner mandate	Refiner must meet fuel specs.	Refiner must meet the more stringent of the emissions cuts tied to fuel specs or to more general performance standards.
Fuel Specs		
Benzene Max vol %	1994: 1	1994: 0.8
Aromatics Max vol %	1992: 30 1994: 28 1996: 25	Same Same Same
Oxygen Min wt %	1992: 2.0 1993: 2.5 1994: 2.7	Same Same Same
Lead	Banned 1/1/92	Banned 1/1/94
Antideposit additives	Required 1/1/94	Same
NOx	No increase allowed	Same
Waiver	Waiver allowed from 2.7 wt % oxygen rule if it interferes with attaining any air quality standard except CO.	No waiver for air quality standard interference. Waivers available for oxygenates as CO strategy.

crude in the United States and elsewhere outside OPEC is likely to ramp up crude prices in the 1990s.

To meet the expected demand for reformulated fuels under new air quality mandates, refiners will have to install additional capacity for distillation, hydrocracking, higher olefins alkylation. They also must undertake sweeping changes in refinery process configurations, including cutting cat cracker gasoline feed differently for conversion to distillates and a whole host of ways to cut olefins in gasoline.

Further, there are the logistical problems that will increase refiners' transportation costs. Reformulated fuels will be a small niche at first. That means segregated shipping and storage in many instances. Pipeline specifications for reformulated fuels will have to be watched closely. Because pipelines moving product from the Gulf Coast to the East Coast would not accept high Rvp gasoline after April, because of new volatility rules effective May 1, shippers scrambled again for supplies in 1990 and some Nymex traders were unable to make physical deliveries on futures contracts.

There are other logistical concerns. For example, what if reformulated fuels are mandated in Los Angeles but not in San Francisco, and the Los Angeles refiners can make only about one-fifth of the reformulated product needed in Los Angeles? Standard gasoline would have to move from Los Angeles refineries to San Francisco and reformulated gasoline would have to move to Los Angeles from San Francisco refineries. That's a double hit on transportation costs.

As U.S. refiners are forced to remove more volatile components from gasoline and rely more heavily on oxygenates, they may be forced to reduce crude runs. Keeping the crude run up would call for making extra jet fuel and fuel oil, thus creating another marketing problem.

Octane will shift among grades as well as oxygenates' share of the gasoline market grows. If a high oxygen content for unleaded is mandated, then there will be excess octane for regular. Conversely, taking some of the components out of premium may squeeze octane for that grade. Thus, components such as alkylates, hitherto considered low value and directed to petrochemical feedstocks, will be prized for their octane capability and relative lack of reactive emissions.

Will all of these required changes also mean a contractor crunch? Installing isomerization, alkylation, or MTBE capacity is costly and sometimes takes several years to permit and build. If a dozen refiners in one region try to revamp at the same time to meet a tight, federally mandated schedule, there may not be the facilities to design and build all this new equipment.

All of this means U.S. refiners must urgently seek cooperation and flexibility from the government on reformulating fuels. Flexible deadlines, as was the case with lead phasedown when refiners could deviate from mileposts so long as they developed their own timetables and stuck to their mileposts and still met the final deadline, are crucial. Refiners should seek temporary waivers to market fuels not yet up to spec, being allowed to pay some economic penalty to equalize the competitive situation, if they lag behind for reasons out of their control.

Flexibility on fuels specs is critical as well. Allowing refiners to tailor their crudes and refinery configurations to meet a general performance standard helps them avoid dependence on a narrow range of feedstock streams and inflexible operations. The alternative may very well be increased imports of refined products.

Other aids to flexibility and timing are bartering and pooling of credits on performance standards, much like the air emission offsets in effect in California today.

Without this flexibility, U.S. refiners will suffer. Independent refiners will suffer the most because they are less likely to be able to afford the facilities needed to supply reformulated fuels. Some could go under. As the U.S. refining infrastructure dwindles, then U.S. producers will suffer as well, because the larger refiners can more readily turn to foreign sources of crude. Generally, independent refiners have been the mainstay of independent producers. So inflexibility on reformulated fuels could contribute to a greater concentration of U.S. upstream and downstream markets to the bigger players.

Given the time and the flexibility, America's refiners should be able to supply the expected demand for the fuels of the future.

Automakers' Role

. . . .

Automakers' efforts individually and in concert with refiners could have signifi-cant repercussions on the consumption and use of motor fuels in the 1990s.

An unprecedented joint research program is under way involving the "Big Three" automakers and 14 oil companies to evaluate reformulated gasolines and methanol fuel blends in current and future vehicles.

The first phase involves studying reformulated gasolines that can be produced and used today. The second phase calls for more advanced research involving new gasoline blends that can be used in future protoype vehicles with advanced emission control systems, as well as testing of methanol and CNG fueled vehicles.

The auto industry is already at work on a new prototype catalytic converter, one that uses electric preheaters and replaces the ceramic honeycomb in a catalytic converter with a metal one. Projections are that this new device alone could cut hydrocarbon emissions by 30%, NOx by 10%, and CO by 60%.

Another auto industry design change is being forced upon it: onboard fuel canisters to reduce hydrocarbon emissions during fueling. The EPA's proposed rulemaking could cut emissions of volatile organic compounds by 5%.

In the longer term, auto industry research could lead to a radical reshaping of transportation in the United States. A research program under way at GM, the Pathfinder Project, seeks to advance the concept of the "smart" car and the "smart" highway. Akin to similar research under way in Europe and Japan, the idea involves using a system of electronic maps and specially equipped vehicles in tandem with two-way communication systems linking the vehicles with a mass transit center. Appropriately, the three-year test is taking place in California.

In the next generation of such technology, predicts GM, there would be an onboard navigation system, a traffic control center, a two-way communications link, and highway sensors to detect traffic densities and vehicle speeds. Such an approach could easily divert motorists from congested routes to less crowded routes. Another possibility is "headway control," whereby an advanced form of cruise control could cut congestion by allowing vehicles to travel safely much closer together.

Such futuristic notions are not that far off, GM insists. As automakers press such efforts and continue to introduce more fuel efficient vehicles, that will tend to dampen the demand for motor fuels beyond simple fleet turnover.

Other Refining Concerns

U.S. refiners face critical environmental challenges beyond pressures on motor fuels.

Principal among them are concerns over emissions of simple pollutants and air toxics, disposal and treatment of hazardous wastes, and treatment of wastewater streams.

Another critical area of concern will focus on plant safety, especially in the light of a rash of deadly accidents in recent years.

Because these problems are essentially shared with petrochemical operations, they will be covered in the next chapter.

Marketers' Concerns

Marketers of petroleum products have their own agenda of concerns regarding environmental issues in the 1990s beyond the repercussions of upstream issues.

The most pressing concern for marketers involves EPA's rule that businesses owning underground storage tanks (UST) are required to have a minimum of $1 million pollution liability insurance. That is not a problem for major oil companies, but more than half the independent marketers in the United States do not have that insurance. Further, pollution insurance rates are likely to continue climbing in the 1990s, making it impossible for many independent marketers to renew their policies because of soaring premium costs.

Hundreds of independent marketers are technically in violation of that rule, which took effect in October 1989. EPA is not pressing the issue now, as long as marketers are making a good faith effort to comply with the UST rule's technical standards (Table 10–5). However, marketers are in a Catch-22 situation, says PMAA.

Table 10–5 EPA's UST Regulations

- All new UST systems must be protected from corrosion, equipped with spill and overfill prevention devices, and furnished with leak detection. Leak detection required on older tanks as of December 29, 1989.

- Existing USTs must have lead detection systems, phased in during a five-year period, starting with oldest tanks. First year compliance deadline was December 28, 1989.

- All existing USTs must be upgraded to new tank standards by 1998.

- Any suspected releases must be investigated. Corrective action requirements for confirmed releases to be determined site-by-site.

- Various requirements for closure, operations, management, reporting, and recordkeeping for the life of a UST.

- State UST regulations take precedent over federal law provided they are at least as stringent.

- Dual financial responsibility of $1 million/occurrence plus an annual aggregate amount.

- For 100 or fewer tanks, $1 million aggregate is the minimum; 101 or more, $2 million; 1,000 or more, $20 million.

- Financial responsibility proof required as of January 1990 for 1,000 or more tanks; October 26, 1990, for 100–999 tanks; April 26, 1991, for 13–99 tanks; and October 26, 1991, for less than 12 tanks.

- Financial responsibility can be shown as: tangible net worth of at least $10 million, plus other requirements, including insurance coverage, a parent firm's guarantee, a surety bond, or a letter of credit; use of an established UST fund or any other method approved by a state; trust fund; any combination of the aforementioned.

According to the association, marketers are unable to make the tank upgrades required by law because banks are unwilling to lend money to marketers for the upgrades over fears of potential environmental liability problems. At the same time, insurance companies are telling marketers that they must make the upgrades before they will be considered for insurance.

What is needed, says PMAA, is a clarification from EPA on what the actual liability is for lending institutions when making loans for tank upgrades. Another help would be establishment of a federal environmental loan program for small businesses to help independent marketers comply with the new regulations.

The UST regulations also call for $2 million minimum pollution liability insurance for companies with 101 to 999 tanks and $20 million for companies with more than 1,000 tanks.

There is legislation pending in Congress that would suspend the deadlines for EPA's financial responsibility requirements for one year. Another bill would clarify the definition of lender liability and require EPA to suspend enforcement where a tank owner submitted evidence that he requested and received a reply for UST insurance.

EPA already took a step toward alleviating the problem by extending its deadline for UST financial responsibility requirements for one year for companies owning less than 100 tanks. That still poses a problem for the many mid-sized independent marketers that own more than 100 but less than 1,000 tanks and don't have the resources to meet the required premiums or can't obtain the financing needed for the mandated upgrades.

Marketers also face a worrisome issue regarding used oil, which may wind up on EPA's list of hazardous waste substances. That will come up again in the 1990s as reauthorization of the Superfund bill is debated.

CAA reauthorization is likely to provide for Stage II vapor recovery controls at service station pumps in certain nonattainment areas. These gasoline vapor recovery nozzles are already widely in use in California, St. Louis, and other areas.

EPA also has proposed a rule that would slash benzene emissions during gasoline handling by 68%.

It estimates the benzene rule could cost $932 million in capital costs and operating costs of $105 million/yr.

The rule does not extend to refueling of vehicles, but does cover about 1,500 bulk terminals, 15,000 bulk plants, and 390,000 gasoline service stations in the United States. Refueling emissions would be covered under Stage II proposals in the CAA legislation.

The benzene rule involves installing a variety of seals, vapor recovery, and vapor balance systems at these facilities.

Upon transfer of properties or closure of service stations, marketers also will have to contend with the costs of cleaning up soil contamination resulting from fuel leaks or surface spills. Solutions include soil excavation and carbon/

platinum catalyst treatment. Another method involves vacuuming gasoline vapors via soil test wells and incineration of those vapors.

A Final Note

. . . .

Soaring environmental mitigation costs could result in the export of much of U.S. refining capacity overseas.

That is the view expressed by Texaco Inc. General Manager of Corporate Planning and Economics, Clement Malin, in testimony on developing a National Energy Strategy before DOE.

"A realistic NES must recognize that foreign export refineries will have an economic advantage in competing for the United States market if their environmental costs are lower, reflecting less stringent emission control and waste disposal requirements," Malin said.

He recommended inclusion of an environmental tax credit for U.S. refiners so that they could boost environmental investment without crippling their ability to compete.

"In the absence of the proposed tax credit—perhaps similar in concept to the research and development tax credit—the United States may well witness the movement abroad of refineries to make transportation fuels, with the consequent impact on the nation's balance of payments and energy security," Malin said.

That is a highly commendable idea and one that the petroleum industry should pursue throughout the breadth of its operations.

It might be worthwhile to extend that logic a bit further, as the United States has often sought to extend moral principles to trade matters, whether it be boycotting South Africa over apartheid or denying the USSR trade privileges over human rights concerns.

Should not the advocates of sweeping environmental reform in this country also insist that the crude oil and petroleum products the United States receive also be searched for only in nonsensitive regions, produced and developed purely, transported safely, and manufactured without effect to the ecosystem? Why not subject the rest of the world petroleum industry to the same stringent environmental standards, either through tariff or other trade incentives (or disincentives)?

If the U.S. environmental lobby wishes to be ecologically correct down the line, should it not be consistent and advocate the extension of standards it proposes for the United States outside this country?

If not, then why does it stand for "exporting" U.S. "pollution" abroad?

References

· ·

1. George Unzelman, Hy Ox Inc., "Ethers in Gasoline—Part 2," *Oil and Gas Journal,* April 17, 1989, p. 44.

11

PETROCHEMICALS AND GAS PROCESSING STRATEGIES

U.S. petrochemical producers face essentially the same challenge as refiner-marketers: Their products are coming increasingly under fire as an environmental problem, and there is a movement under way to find substitutes thought to be more ecologically benign.

Principally the concern over petrochemicals centers on two areas: the garbage glut facing this country and chemophobia, or the Bhopal/Love Canal syndrome.

In both areas, the solution is essentially the same: Industry must establish zero as a goal to strive for: zero refuse through recycling, zero emissions, zero hazardous waste, zero discharge, zero risk.

Although the zero sum game is an impossible ideal if any industrial enterprise is to survive economically, the ideal should form the basis of a fundamental change in industry practices.

In petrochemical operations, the industry has long been regulated under the theory of remediation. That entails reducing or cleaning up an existing environmental contamination.

It is time now to focus exclusively on preventive regulation, intended to keep a contamination from occurring in the first place.

Minimization must be the watchword for the petrochemical industry. There is no industry at greater risk for being phased out of existence in many market areas if the byproducts and end products of its manufacturing processes are not on a track toward elimination.

There are no practical, economic substitutes for oil and gas in their primary markets. There is, however, a wide range of substances, old and new, that can be substituted for many petrochemical products, especially plastics. It cannot be forgotten that the petrochemical industry essentially was born of a need to substitute strategic materials on a crash basis during wartime.

The explosive growth of the petrochemicals industry since World War II—and especially in the past 20 years—owes to technological advances leading to better substitutes for conventional materials.

The petrochemicals industry now is embarking on a new era of specialty plastics, engineered polymers, and other products that are making inroads into markets hitherto dominated for generations by metals, woods, glass, and other materials.

But the industry could face the loss of its bread and butter markets if there are no adequate solutions to environmental concerns.

Perhaps because of the advanced state and ubiquitous nature of chemophobia in U.S. culture, no other industry has done more, especially in the past few years to speak to environmental concerns—out of necessity.

Companies like Dow, Du Pont, Monsanto, and Union Carbide have been among the leaders in American industry at seizing the initiative on environmental concerns, as has been described elsewhere in this book. There is much more still to be done.

As for gas processors, many of the environmental rules and laws and trends applying to refiners and petrochemical producers also apply to them. It is the nature of gas processors' principal products' role as feedstock for refined and petrochemical products that their fortunes are inextricably intertwined.

Solid Waste Problems

· ·

Every American throws out more than four pounds of garbage each day. That adds up to more than 150 million tons of municipal waste generated in the United States each year.

Because waste management practices have lagged the explosion of new consumer goods and packaging in the marketplace, in addition to the pressures from a growing population, the United States is experiencing a garbage crisis.

All but about 20% of the nation's waste goes into landfills. However, we are seeing the limits of existing landfill capacity. That is due in part to the closure of many landfills—more than 12,000 out of 18,000 in the past 20 years—in part because of the degradation they have caused to the surrounding environment. Waste management experts predict half of the remaining landfills will close in the early 1990s. Problems encountered include buildup of toxic gases and leaching of toxic liquids into groundwater.

Further, it is a virtual impossibility to site a new landfill in the United States today, despite the growing crisis. Aside from the poor environmental track record of landfills in the past, most communities simply do not want them for quality of life/aesthetic reasons.

Why all of this poses a major problem for petrochemical producers is that the environmental lobby, aided by government, has singled out plastics as a Public Enemy No. 1 in the solid waste debate. Notably targeted are the plastic foam (usually polystyrene foam) packaging and other items in food service. Because these foam boxes, cups, and trays are as ubiquitous as the hundreds of thousands of McDonald's and Burger King, et al., restaurants from whence they came, they are especially glaring examples of litter. Further, foams have been implicated in the purported destruction of the ozone layer because chlorofluorocarbons have been used as blowing agents in the manufacture of polystyrene foams.

Doubly villainous to environmentalists, polystyrene foams have invited municipal bans, taxes, and ordinances across the country. That campaign has since broadened into an all-out assault on all plastics, usually with the dire predictions of a typical plastic container taking 500 years to decompose. That scenario has always puzzled. Plastics have not even been around 100 years yet. How would they know?

None of the disposal solutions are panaceas. Even well-operated landfills are not necessarily welcome by communities. And the increasing cost of maintaining and operating municipal landfills, especially with growing regulatory intervention by the government, is making that even less of an economic proposition.

In New Jersey, for example, where the state government has been especially aggressive in regulating the waste disposal industry like utilities, complete with rate regulation, the number of waste haulers has fallen to about 650 from more than 2,300.

Another promising solution, waste incineration, preferably in the form of resource recovery plants, has been stymied by permitting and economic concerns. Other trash-to-energy projects involve extracting methane from biomass. These projects have established a good track record in converting waste to energy, but the slide in oil prices in the 1980s hampered their economics in many areas. In addition, tightening air quality rules threaten to impose huge mitigation costs. In many areas, resource recovery plants are economic only because the costs of landfilling have soared out of sight and the power they generate receives incentive prices under federal or state law.

Plastics' Role in Municipal Waste

For all the hype, plastics are getting something of a bum rap in the solid waste debate.

According to some environmentalists, plastics may represent as much as one-third of the U.S. waste stream by volume. What they don't acknowledge in these estimates is that such numbers relate to uncompacted waste, before plastics are compacted in a garbage truck and further crushed at the landfill.

According to EPA, plastics comprise only 7.3% by weight of the nation's total waste. Of that amount, plastic packaging—singled out by environmentalists—accounts for only half. As for the "doubly evil" plastic foams used in food service and food packaging, they account for only 0.25% of the nation's total solid waste stream. Furthermore, almost all of the major manufacturers of polystyrene foams have dropped use of CFCs as blowing agents in their manufacture.

For all of that, then, why do almost eight out of 10 Americans—according to a 1989 industry poll—see plastics as a serious environmental threat? Again, the ascendancy of hype over scientific fact rears its head.

The ire over plastics has to do with their nonbiodegradability. Scientists have no consensus on how to predict or show biodegradability, so the public is left with the environmental lobby's unsubstantiated and often lurid claims.

In fact, nondegradability of plastics is actually regarded as something of a plus for modern landfill operations. The environmental problems that have occurred at landfills have resulted from: too-rapid decomposition of some materials, resulting in buildup of noxious or explosive gases; longer term decomposition eventually creating land shifts such as sinkholes; and decomposing materials, such as inked paper, leaching contaminants into the groundwater.

Modern landfill practice calls for a design that minimizes biodegradability for that reason. Having nondegradable plastics in a landfill actually lends some stability to the landfill. It is worth pointing out that modern landfills usually are designed with plastic liners to help prevent leaching out of toxic substances and isolating them for later removal. So plastics' contribution to the landfill capacity problem is a red herring.

In fact, biodegradability may itself be a myth, according to the results of some pioneering work by a University of Arizona archaeologist, William Rathje, who is conducting landmark research on landfills. Rathje, upon learning that no one had ever dug into a landfill to find what was inside and what was happening, set about to do just that.

In 1987, teams led by Rathje excavated three municipal landfills—in Tucson, Chicago's suburbs, and the San Francisco area—all of which had still been receiving trash during 1977–85.

In 1988, a Rathje team excavated another landfill in the Chicago area, which had been active during 1970–74.

In the 1987 digs, Rathje found that newspapers, paper bags, telephone books, grass clippings, and foods such as meats and vegetables were just about as well preserved as the plastics. In the 1988 excavation, such organic debris as grass clippings, hot dogs, and a T-bone steak still with lean and fat showing, was in even better preserved condition than in the earlier digs.

Incidentally, Rathje's work also refuted some claims that plastics take up a disproportionate volume of space in landfills in proportion to their weight. He found plastic bottles crushed just as flat as other containers.

The upshot of this work, according to Rathje in a 1988 speech before the annual meeting of Keep America Beautiful Inc. is that "the major conclusion we have reached after our landfill excavations is that under current management practices, the compost theory does not work."

Even with that body of evidence, petrochemical producers have ventured into an ill-advised attempt to market biodegradable plastics products.

Noteworthy among these were Mobil's introduction of biodegradable trash and grocery bags and Coca-Cola's switch to photodegradable materials in plastic ring connectors in six-packs and in plastic bottles.

An environmentalist fusillade against Mobil led the company to drop its program labeling grocery and trash bags, justifiably citing Rathje's work. (It is ironic, however, that their previous fervent adherence to the doctrine of biodegradability led to the consumer pressures on Mobil to introduce the product in the first place.)

As for the concept of photodegradability in plastics, environmentalists quickly pointed out that the concept was useless in landfills, where light is absent.

They also raised questions about the plastic residue's effects on the ecosystem after the degradable components in the plastic—generally derivatives of corn starch—have decomposed. Their most compelling argument, however, was that photodegradability may actually encourage people to litter.

There are questions as to whether composting efforts involving aeration, moisture, and micro-organism enrichment might aid in decomposition of degradable plastics.

Even if there are advances that might make biodegradable plastics truly feasible, one inevitably runs again into the problems of landfill space and litter.

The industry should learn from the early debacles involving widescale introduction of a product labeled biodegradable. Face it: The industry hardly has a monopoly on credibility in the public's eye. Simple efforts to cash in on a marketing trend are not going to work. There really is only one solution that makes sense, and that is the solution that has worked for the glass, aluminum, and paper industries: recycling.

Recycling Push

· ·

The petrochemical industry is just beginning to test the waters of recycling its products (Table 11–1).

It should greatly accelerate that effort, because the alternative is to see many of its products regulated or legislated out of existence. With another summer like that of 1988, with garbage barges wandering the seven seas like some Flying Dutchman of refuse and dangerous trash washing ashore on beaches, the end result could be an epidemic of ordinances banning plastics or mandating their recyclable substitutes. Why wait?

If loss of markets is not enough incentive, there are sufficient other reasons from a sound business practice perspective to recycle.

Recycling is an economic proposition for plastic producers as well as for communities. In some areas, recycling plastics runs about $50/ton compared with as much as $100/ton for land-filling or incineration.

Producers save on feedstock costs for some products, manufacturing some goods from recycled products for about half the cost of virgin plastics, according to data from the Plastics Recycling Foundation (PRF).

Table 11–1 United States Plastics Recycling Programs

Company/Venture	Location/Start-up	Product	Volume
Amoco	Brooklyn, 1989	Polystyrene foam	N/A
Clean Tech Inc.	Dundee, Michigan	PET, HDPE bottles	N/A
Du Pont, Waste Management Inc.	Philadelphia, 1990	Various	40 million lb/year
Du Pont, Nat'l AgChem Ass'n	Findlay, Ohio, Washington Company, Mississippi, 1989	HDPE pesticide containers	40,000 lb/year (pilot program)
National Polystyrene Recycling Co.*	New England, 1990, others, 1990	Polystyrene foam	250 million lb/year by 1995, 25% of all
Occidental	Berwyn, Pennsylvania, 1989	PVC bottles	Under development
Resource Recycling Technologies Inc.	Binghamton, New York (Various, also markets recycled products)		
Sonoco, Graham, Exxon, Valvoline	York, Pennsylvania, 1990	PET milk bottles recycled	100 million motor oil, auto fluids bottles

* Formed by Amoco, ARCO, Chevron, Dow, Fina, Huntsman, Mobil, and Polysar. First stage involves McDonald's restaurants in New England. Recycling centers also planned by yearend 1990 in the San Francisco Bay area, Los Angeles basin, Chicago, and Philadelphia.

Assuming a recycling plant producing 20 million lb/yr of recycled product, PRF calculates cost of capital at about 2.5¢–15¢/lb and feedstock cost at about 5¢–6¢/lb. After processing, depreciation, and return on investment, the recycled product can be manufactured at a cost of about 15¢–30¢/lb.

For a comparable virgin plastics plant, the cost of capital would be anywhere from 20¢ to $1/lb and ethylene feedstock costs about 30¢–35¢/lb. After depreciation and ROI, the virgin product costs about 40¢–60¢/lb to produce. Those figures were calculated in mid-1989. Although ethylene prices dipped sharply from that level in 1990, projections of rising oil prices are likely to pull ethylene prices up as well, and with that the economics of plastics recycling.

PRF contends the quality comparison of virgin versus recycled between paper and plastic is another plus. Even the Sierra Club has to swallow hard in putting out its beautiful coffee table books and calendars on virgin paper because recycled paper just does not cut it when it comes to high quality photographic reproductions. But with plastics, it is difficult to tell the difference between virgin and recycled, PRF contends. Some of these recycled plastic products may even provide petrochemical companies entry into new markets, such as construction and agricultural materials.

There is a downside to recycling for petrochemical producers, however. Widespread recycling at a time when the petrochemical industry has been frenziedly adding capacity may contribute to the overbuilt capacity and soft markets that many analysts are predicting in the mid-1990s.

Chem Systems Inc., Tarrytown, New Jersey, in a 1989 study, extrapolated from EPA's target of 25% solid waste recycling by 1992 a projection of as much as 5 billion lb/yr. But the analyst sees that as unlikely without a lot of government intervention. Its own estimate is that recycled plastics volume will reach 1.1 billion lb/yr by 1992. Of that total, high density polyethylene recycling will account for about 560 million lb—a tenfold increase from its level in 1987.

With soft drink bottle recycling reaching a likely level of 50% by the mid-1990s, recycled polyethylene terephthalate will total about 600 million lb/yr, Chem Systems projects. That translates to the equivalent of about 1 billion lb/yr of ethylene, or a world-scale steam cracker.

With companies already pulling back from announced plans to add ethylene and downstream capacity in the 1990s, it would behoove petrochemical producers to factor the effect of recycling on demand to their plans for the 1990s. It could prove significant enough to warrant postponing or even canceling planned expansion projects.

But recycling's dampening effect on demand should not be read as a reason to hesitate about plunging into it. The economics, new market opportunities, and community goodwill aside, the real concern is the alternative: a widespread banning of plastics. That is a very real threat. Now is not the time to be complacent.

Hazardous Waste

• •

If there is some uncertainty about how clean air and solid waste concerns will affect refiners and petrochemical producers, respectively, there is little doubt about the effect of federal legislation on hazardous waste for all process plant operators:

It is coming, perhaps sooner than processors care to acknowledge; it will be very, very costly.

Processors have a variety of options to deal with the increasingly stringent regulatory treatment of hazardous waste under federal and state law. None is a magic solution, all will probably be used to some degree, and there is a growing likelihood that industry's capacity to handle the hazardous waste it generates will be stretched to the breaking point—the point at which some facilities may simply have to shut down until solutions are found.

Disposing of and treating hazardous wastes is regulated under the 1976 Resource Conservation and Recovery Act (RCRA). Essentially RCRA's focus is preventive. It was designed to prevent process plant operators from disposing of hazardous wastes unless those wastes first were treated to nonhazardous status. RCRA rules determine methods for treating, storing, and disposing of hazardous wastes. They also mandate that all such wastes be monitored from "cradle to grave." The idea is to avoid situations such as that which led to the creation of the Superfund program (Comprehensive Environmental Response, Compensation, and Liability Act). Superfund covers hazardous waste generated before November 19, 1980 (RCRA thereafter).

In 1984, approval of the Hazardous and Solid Wastes Amendments (HSWA) was designed to ensure that EPA and other federal environmental regulatory bodies act to improve RCRA's hazardous waste provisions. What HSWA especially focuses on is eventual elimination of all land disposal of hazardous wastes, setting deadlines for certain specific wastes. If EPA does not override a land disposal ban for a cited waste by the deadline, an automatic total ban goes into effect under HSWA's "hammer" provisions. The hammer provisions cover facilities that generate, treat, transport, store, and dispose of hazardous wastes.

EPA classifies hazardous wastes according to toxicity, reactivity, ignitability, or corrosiveness. It also has the authority to add substances to its current lists. Complicating processing operations' handling of hazardous waste are EPA's "derived-from" rules. These cover mixtures and hazardous and nonhazardous solids wastes.

All such wastes are subject to hammer provisions and eventual land disposal bans. Under HSWA, disposal involves treating substances in order to render them nonhazardous. The standard for treatment under HSWA is the best demonstrated available technology (BDAT). After August 1990, any wastes treatable by BDAT will be banned from land disposal unless EPA specifically exempts them case-by-case. Such exemptions could occur if a plant operator can guarantee that land disposal will not result in migration of the hazardous waste through runoff, groundwater leaching, etc.

By the end of summer 1990, the variances that refiners have had for disposing of hazardous waste solids and sludges will have disappeared. A critical deadline passed in May 1990 that mandated treatment standards for more than 415 hazardous wastes, in addition to the 121 wastes covered by land ban standards since 1986. Any leeway EPA had in implementing those standards (e.g., "soft hammers") for the most part died in May 1990.

For refiners, that leeway extended a few months into August 1990 for dissolved air flotation float, slop oil emulsions, heat exchanger sludge, API separator sludge, and tank bottoms.

EPA probably will publish the final treatment standards for hazardous wastes on its lists in late fall 1990.

For the most part, the new standards will favor incineration for disposing of hazardous wastes. Disposal concerns won't end there, because further treatment or disposal will be required for resulting ash and scrubber water from incineration.

Refiners can tackle disposal of hazardous wastes in a number of ways. RCRA allows a refiner to dewater hazardous solids, blend them with recovered slop oil, and process the blend in fluid cokers.

Another approach calls for filter pressing, solvent washing, and thermal desorption to treat oily sludges. Oily waste sludges also can be centrifuged after initial treatment to further separate oil, water, and solids. These solids can be pressed into dry filter cake and drummed for disposal or incineration.

Incineration will be the way to go for most processing plants that do not have a land disposal option. Under EPA statutes, incinerators are deemed BDAT if they can destroy 99.99% of the hazardous materials burned.

Recycling will be the preferred option for process plant operators with other materials, notably spent catalysts. Metals from spent catalysts can be reclaimed for sale and remaining materials sold or disposed under RCRA.

As attractive an option as incineration may be, there are drawbacks. Local community opposition is making on-site incineration permits more difficult to obtain. Commercial incineration programs can also run into permitting problems and are much more costly than land disposal.

Because of those circumstances there is reason to believe that there will not be available the capacity for waste treatment in the early 1990s, especially for incineration.

Further complicating the issue is the question of on-site remediation for facilities that seek to gain permits for treatment and disposal of hazardous wastes under HSWA. Under RCRA, a processor seeking a permit for a facility must first deal with any past or current waste problems at that facility. EPA is even stepping up enforcement of remediation at facilities that have yet to receive a permit.

That will squeeze the land disposal capacity situation even further. According to EPA estimates, more than 5,000 waste disposal sites could face corrective action under RCRA. Some could even involve Superfund cleanup.

The costs will be staggering. For the chemical industry alone, RCRA corrective action costs will range $16–$27 billion, according to a study by the Chemical Manufacturers Association (CMA). If EPA mandates incineration for all hazardous wastes, the cost could be as much as $39–$65 billion, according to the CMA study.

The amount of hazardous waste covered by RCRA could expand even further if newly proposed EPA rules governing residues and wastewater are put into effect.

What all this portends is that EPA is inching toward a zero standard for the amount of any hazardous substance in a waste stream. Even the most microscopic trace of a hazardous substance in a waste stream could ultimately render that entire waste stream subject to BDAT treatment. The processing industry needs to press for RCRA exemptions for waste streams with hazardous contaminants where the hazard is truly negligible. That way, EPA can devote its resources more properly to the truly significant hazards. Without such exclusions, the processing industry is almost certain to run up against a disposal capacity crunch.

Industry can expect this tightness to worsen after EPA issues its final regulations governing on-site industrial boilers and furnaces that burn hazardous wastes for their heat value. If the same stringent criteria for polluting air emissions and toxic air emissions that apply to permitted hazardous waste incinerators are also applied to these industrial boilers and furnaces, then much of the practice will be curtailed for economic reasons. The idea is that the permitting process itself could prove so costly as to outweigh the savings in fuel from keeping the boilers operating. The rule could affect more than 1,000 facilities.

The hazardous waste disposal issue could be exacerbated by regional discrepancies. In areas where there is a shortfall of such capacity, hazardous waste generators are obliged to export the wastes for landfill or incineration. However, states or regions with surplus capacity are worried about becoming dumping grounds for hazardous waste, not to mention the increased risk of transporting these substances across their borders.

The best approach for process plant operators to reduce the problems associated with treating, handling, storing, and disposing of hazardous waste is to minimize its generation at the source.

This should involve a comprehensive survey of every substance existing at the plant at each stage of its existence, and a program to put each substance out of the hazardous waste equation. One area that will attract intensified application by the process industries is membrane separation technology.

Ultimately, whatever standard comes out of proposed EPA rulemakings or congressional legislation is one that will establish either a broad or a specific target for all hazardous waste generators to reduce the volume of waste generated to a specific percentage of a plant's total product yield. That won't work for some plants or some materials, which will probably mean another series of EPA "soft hammers" and exemptions that the agency and industry will continue to grapple with well into the next century.

Waste Minimization

. . . .

Ideally, the best approach to waste minimization is one that strives for a zero standard. By reconfiguring process approaches, a refiner or petrochemical producer may find a way to eliminate a hazardous waste stream within the processing to avoid costly end-of-pipe solutions. As the costs of waste treatment and disposal continue to escalate, process plant operators accordingly must consider process changes to eliminate hazardous waste into the economics of their operations.

Companies are finding a side benefit of significant cost savings in such approaches and staying vigilant of new ways to reduce generation of hazardous waste still further. For example, Monsanto long has had a strong waste minimization program. But the curve of operating results from the program had begun to flatten out. Previously, Monsanto had concentrated on a single waste stream at a time. The new approach called for a variety of strategies to reduce chemical components in all waste streams. Further, Monsanto established quotas for plants to achieve a specific percentage cut in the targeted chemicals across the spectrum of waste streams. To follow that up, the company is revamping its database to reflect the new broad-brush approach.

Some companies have pushed waste minimization pretty much to the limit, they think. But it behooves process plant operators to rethink what the limits are for waste minimization—and not just apply loose or outmoded cost-benefit standards—and continue to try to "push the envelope." Process plant operators who develop new technologies in this course could also create new profit centers and market a technology that should be in great demand.

The best incentive for aggressively pursuing waste minimization is heading off new legislation on federal hazardous waste reduction standards—say, a 95% universal standard for all plants dealing with hazardous waste. Reducing waste to 5% of a chemical's output could be the focus of RCRA reauthorization legislation in the early 1990s.

At the same time, refiners and petrochemical producers should assert themselves and remind the public, as in the previous chapter, that extreme mitigation measures carry heavy costs that are not shared by foreign competitors. Unless there is effort to deal with this disparity, American competitiveness will suffer.

That issue will come up again in 1991 as Congress considers reauthorization of the Superfund law. The Superfund program has cleaned up only a fraction of the 1,200 or so sites on the National Priorities List (NPL) at an average cost of $200 million/site.

Because of the slow progress, EPA wants to shift cleanup costs for some sites onto the states or defer to RCRA for other sites. Environmentalists oppose this approach, claiming sites removed from the NPL will not undergo the extensive cleanup required under Superfund. There is a concern for industry that such an approach will involve a widely disparate approach among states to cleanup. A happy medium might be a states/RCRA scheme if administered jointly through EPA regional offices and state agencies. When this happens, it will be more imperative than ever that the petroleum industry make its case on the issue of equitable cost-sharing for orphan dump sites, so it does not continue to shoulder an unfair burden under the Superfund tax.

Not that the RCRA approach will be cheap. Under standards and procedures expected to be published in the Federal Register in late summer 1990, EPA estimated as many as 4,000 facilities will conduct hazardous waste cleanups at costs ranging from $10 to $60 billion. The EPA to begin conducting hearings on the new standards in October 1990.

Air Toxics Concerns

· ·

Air pollution issues will remain near the top of the environmental agenda for process plants in the 1990s.

Concerns over smog, global warming, ozone layer degradation, and acid rain related to new legislation and regulation have already been covered here.

What is especially worrisome for processors is the new regime of law and rule that will govern the emissions of air toxics. Resurrecting the specter of Bhopal is enough to convey to the public a more immediate danger than fuzzy speculation over phenomena that may or may not occur decades hence.

The framework for federal regulation governing air toxics was put forth under provisions of Title III of the 1986 Superfund Amendments and Reauthorization Act (SARA)—the Emergency Planning and Community Right-to-Know Act. SARA Title III requires all industries—but most notably refiners and chemical producers—to disclose inventories of chemicals on-site and their relative emissions. It also obliges companies to develop emergency contingency plans together with local communities.

In spring 1989, EPA issued the first report estimating the amount of toxic chemical releases to the environment based on the SARA Title III reporting requirements. When Rep. Henry Waxman (D.-Calif.) released the portion of the EPA report dealing with air toxics via press conference and press release, it generated widespread publicity and public concern. Playing to the public's love of lists and rankings, Waxman held up certain specific plants or companies as the worst polluters without regard to a scientific assessment of whether these facilities actually constituted any threat to public health or safety. Several companies winning the unhappy designation of No. 1 on this list scrambled to implement public relations damage control, with the upshot being that Waxman had used data that were at least two years—and sometimes more—out of date, referred to practices no longer in effect, and sometimes just got the information wrong altogether.

Here is the real danger of SARA Title III to the processing industries: that the data will be used by environmental lobbyists and their congressional proxies to further their political agenda at the expense of facilities that pose no significant, unnecessary risk to the public.

Until now, state and local agencies have taken the lead on regulating air toxics. With reauthorization of the CAA, however, the EPA will become the dominant force.

Under CAA legislation, the standard for determining whether a facility is considered a major source of toxic air emissions will be tightened considerably. The Bush administration's CAA bill called for a standard of 10 metric tons/yr for a single pollutant on the list or 25 tons/yr of combined listed pollutants. The current standard is 100 metric tons/yr.

Air toxics will be regulated under a permit system run by the states. A major concern will be facilities that produce a varying slate of products at differing times—which pretty well covers most refineries and petrochemical plants.

CAA air toxics rules are likely to toughen the standard for emissions control technology. Currently, that is Best Available Control Technology (BACT). The new standard probably will be Maximum Achievable Control Technology (MACT).

According to the Bush legislation, MACT is defined as the best emission control achieved in practice by the best controlled similar source. MACT will take into consideration economic, environmental, and energy use effects. It will be up to EPA, however, to interpret the definitions on a case-by-case basis. The major point of concern for industry is to what degree EPA yields to the environmental lobby's opposition to applying cost-benefit analysis to MACT standards.

Further, the MACT standards won't necessarily stay the same. Seven years after enactment, the CAA legislation can reopen assessment of the standards to evaluate public health risk, then mandate even tougher standards if deemed necessary.

Permits granting a variance from MACT would be allowed if a source can show that alternative controls pose a negligible risk to public health or the facility has achieved a 90% cut of all listed pollutants or 95% cut in listed particulate pollutants within five years.

Congressional CAA legislation features other issues of concern to the process industry on air toxics. Especially worrisome is the requirement that EPA develop a list of at least 100 substances known or suspected to be a threat to human health. EPA then would develop regulations governing these substances under the circumstance of an accidental release and require facilities to develop risk management plans. Facilities operators would have to install costly new equipment aimed at preventing and controlling these releases and train their person nel accordingly. The concern is one of degree. One House bill—Rep. Waxman's—would establish a standard of a 1-in-1 million cancer risk.

Whatever the final form of CAA legislation, processing plant operators must take steps now to improve their ability to measure emissions accurately. That way, they can be prepared to outline the effect of the final legislation. At the same time, companies must begin efforts to reduce those emissions wherever feasible in the absence of standards. Every processing plant should already have a risk management plan in place. Not having a federal rule in place won't shield a company from legal liability in the event of an accidental release—and having one may even limit future financial liabilities.

Plant Safety

· ·

Another critical area of concern for refiners, petrochemical producers, and gas processors is the growing alarm over process plant safety in the United States.

That concern probably peaked with the October 1989 explosion at Phillips 66 Co.'s Pasadena, Texas, polyethylene plant, which leveled the facility and claimed 23 lives. That concern was sustained into spring 1990 with reports, claims, counterclaims, and fines related to the accident. Just as OSHA was about to publish its new rules governing chemical process plant safety, a July 5, 1990, explosion and fire in the utilities area of ARCO Chemical Co.'s Channelview styrene monomer/propylene oxide plant claimed 17 lives and effectively shut down that plant.

The questions that have arisen over those accidents and a score of smaller ones in recent years tend to come back to one principal area of concern: Has the petroleum industry increased the risk of accidents by cutting personnel in the lean years following the 1986 oil price collapse? Are process plant operators jeopardizing safety in their plants by relying on inadequately trained contractor personnel?

Central to the investigation of the Phillips plant disaster is Phillips' contention that long established procedures were not followed—and that, apparently, contractor personnel were directly involved. Phillips has been cited by OSHA for safety violations for each employee at the plant, and it is contesting those violations.

278

Whatever the outcome, companies must redouble efforts at indoctrination and training in safety and emergency response and ensure that the standards developed for contractor and subcontractor personnel be just as strict and monitored just as closely as for their own employees.

It is likely that the 1990s will see the evolution of a national board akin to the National Transportation Safety Board set up to regulate, oversee, enforce, and investigate safety practices and accidents in process plants, beyond the scope of OSHA. There already is legislation before the Congress to that effect. Setting up a parallel committee across management-worker-union lines should be imperative for every process plant.

Contractors' Role in Safety

. . . .

It is perhaps a little too simplistic to jump to the conclusion that increasing use of contractor personnel as a cost-cutting measure at refineries and petrochemical plants is related to the string of accidents at such facilities during 1987–90 (beginning with the deadly blast at Shell's Norco, Louisiana, complex).

A study by John Gray Institute of Lamar University suggests an inherent contradiction in assumptions over contractor employees' roles in plant safety. OSHA commissioned the study and will release the final report probably in November 1990.

The John Gray study looked at plant safety practices in nine case studies involving petrochemical plants or refineries and a national survey of process plant managers. Although the study's scope will be broadened, preliminary findings yield some fascinating data.

The survey found that work deferred to contractors increased across the board, but primarily due to industry expansion rather than substitution for permanent employees. Further, contractor usage has increased in mainly traditional contractor areas for reasons of cost, flexibility, specialty in asbestos removal, and capital expansion.

While one-fourth of contract employees did not know how many fatalities occurred among contract employees working in their plants, only 1.4% did not know that number among permanent employees. However, fatalities per million hours worked decline as total hours worked increased—but at the same basic rate for contract and permanent employees. Although accident rates tended to increase with hours worked and plant size, as contract laborers worked more, their injury rates fell.

The survey also indicated that plant managers tend to be reactive rather than preventive in their common approach to monitoring contractor safety: requiring reports of all accidents.

In the case studies reviewed by John Gray, cost and flexibility also drove contractor employment increases. Some plants had no safety criteria for contractors and even some of those with rigorous screening criteria routinely deviated from them. Although most of the plants did not track contractor safety performance, at least two of them replaced contractors because of poor safety performance.

Where there were data, short-term and contract employees had higher-illness and injury rates than did permanent employees (which seems to contradict the comparable survey results). There were other findings related to problem areas: union/nonunion animosity, sharp differences between contract personnel and permanent employees on safety training and approaches, and language barriers on the Gulf and West coasts.

What all of this suggests is not entirely clear. However, common sense would dictate that it be advisable to take the same approach and enforce the same rigorous standards on safety training and awareness for contractor personnel as for permanent staff. It also says that establishing a sense of continuity and consistency among employees of both camps presages a better safety record.

That can be seen in the case of a "model" plant that the John Gray study focused on.

In this model plant, management had almost eliminated distinctions between safety and health practices of contract and permanent employees. Its main goal is to achieve a long-term continuity, figuring that greater familiarity with the plant will result in improved safety and efficiency. Consequently, product quality, productivity, and profitability are exceptional at the plant, the study found.

To work at the model plant, contractors must meet stringent pre-bid requirements, including comprehensive safety training and employee benefit packages. In addition, contract employee wages are commensurate with those of permanent employees.

Upon entering the plant, contract employees are trained and often supervised by a select cadre of plant personnel, and they continue to participate in safety programs. Both the principal contractor employees and permanent workers reported the lowest OSHA injury and illness rates within the study group.

The levels of satisfaction and loyalty among contract as well as permanent employees at the plant are so high that contract workers often have left a former contractor in order to continue working there. Almost 90% of the plant's contract employees have worked at the plant since it was built.

That's a "model" plant. Perhaps that is the model for safety that all process plants should strive for.

New Approaches to Safety

It is time for process plant operators to rethink their basic approach to safety.

A hint of what might be in the offing on safety issues is what Scaqmd has proposed on the issue of hydrofluoric acid at refineries. Scaqmd in April 1990 voted to ban HF acid at four refineries and a chemical plant in Southern California. The ban is to take effect December 31, 1994.

Scaqmd took its cue from a local initiative in Torrance, California, where Mobil was faced with a negative image after a series of accidents at its Torrance refinery—one involving HF acid buildup. The initiative sought to ban use of HF acid at the plant. Mobil overcame huge odds to defeat the measure, which will be covered in the next chapter.

Mobil dodged a $100-million bullet—the cost of switching from HF acid to sulfuric acid at the refinery—with that effort.

Actions by states or communities aimed at facilities they deem to be unsafe and a threat to its neighbors will become more commonplace in the 1990s and take on added impetus as more accidents occur. More accidents are probably inevitable, given the continuing strain on industries pressed to the maximum limit on capacity. Although there may be some relief in sight on capacity for some petrochemicals, other petrochemical products and general refining capacities will remain tight for much of the early 1990s.

With facilities running flat out, companies may be reluctant to shut them down while margins are high. And some critics have accused the refining and petrochemical industries of skimping on maintenance.

Is that logical? Would a process plant operator be so short-sighted as to risk an unscheduled outage because of an accident or equipment failure? Are companies endangering their facilities through cost-cutting measures spurred by intense competition from abroad? If there is one area where industry should aggressively pursue the ideal of zero risk, it is here. Process plants simply cannot afford to take unnecessary chances, if they wish to remain in business.

Union Carbide Chairman Robert Kennedy said, "The most imposing threat to the viability of chemical companies, and perhaps to our very existence as an industry, is the deep concern of the public over the safety of our products and operations."

Kennedy promotes a chemical industry initiative aimed at improving performance in health, safety, and environmental affairs. The initiative calls for establishing a set of guiding principles; a set of management practice codes; a public advisory panel to critique codes and help identify public concerns; a self-assessment program that requires companies to chart their progress in implementing the codes; and executive leadership groups to give member companies the chance to raise new questions and share their experience and ideas.

"We have got to convince a skeptical public that we have the message on health, safety, and the environment," Kennedy said. "The key is no mystery; it is simply real and continual improvement in industry performance."

New OSHA Standards

· ·

Much of industry's new approach to safety will be determined by its fealty to a new safety standard OSHA has proposed for all processing industries covering about 28,000 plants and employing about 2.2 million workers involved in hazardous operations (Table 11–2).

According to OSHA, the new standard will prevent about 200 fatalities and 720 injuries and illnesses each year. It will cost all targeted industries $637.7 million/yr to comply and garner them $404.5 million/yr in economic benefits, for a net compliance cost of about $230 million/yr, according to OSHA estimates. OSHA maintains, however, that its estimates of economic benefits are probably too low because they don't take into consideration such ancillary costs as insurance, administrative overhead, and productivity improvement.

OSHA was to hold a hearing November 27 in Washington, D.C., on the issues raised by the proposed standards.

Table 11–2 OSHA's Proposed New Process Plant Safety Standard

- Process safety management (PSM) program requiring operators to conduct process hazard analyses of their plants.
- Compilation of process safety information and data about levels of toxicity, permissible exposure, reactivity, corrosivity, and thermal and chemical stability of on-site materials.
- Maintenance and effective communication to employees of safe procedures for process-related tasks, including start-ups, special hazards, normal operation, temporary operation, and emergency shutdown.
- Training workers to help them understand the nature and causes of problems arising from process operations and increasing their awareness of specific hazards.
- Ensuring contractors work safely on-site and that their employees are properly trained and informed of potential for fire, explosion, or toxic releases related to their work and the process in addition to general safety rules and emergency procedures.
- Performing periodic inspections of equipment at acceptable intervals and maintaining mechanical integrity of pressure vessels, storage tanks, piping systems, relief and vent systems and devices, emergency shutdown systems, and controls, alarms, and interlocks.
- Issuing "hot work" permits to make sure plant operators are aware of potentially hazardous work under way and that appropriate safety precautions have been taken before work begins.
- Safe management of changes to process chemicals, technology, and equipment by establishing and maintaining written procedures before implementing changes and by notifying and training employees as soon as practicable about changes to come.
- Using a team of experts to investigate potentially major incidents to identify the chain of events leading to them and determine causes.
- Developing emergency action plans telling employees how to handle emergency shutdowns, evacuations, emergency response notifications, notification of other employees, and procedures for controlling emergencies.
- Conducting compliance safety audits at least every three years to assess effectiveness of the PSM in place.

What the new standard amounts to is a codification into federal law of the optimum plant safety practices largely in effect at the best plants. Those practices stem from generally accepted engineering practices or existing safety and health codes. The new standard would give those practices and codes the force of federal law.

What it means for process plant operators is a surge in paperwork, record-keeping, and a new layer of federal regulatory oversight. Most importantly, it means that violating these practices results in a violation of federal law, with the fines, penalties, and perhaps criminal prosecution to follow.

Since the ARCO Channelview blast occurred about the time the new standard was being readied for publication in the Federal Register, the accident clearly did not play a role in its formulation. However, as this book went to press, OSHA was considering ways of suggesting industry improve plant safety during the comment period. Should the string of petrochemical/refinery accidents continue in the fall, look for Congress or the administration to seek emergency legislation giving OSHA statutory authority to implement interim measures.

Changes Ahead for Gas Processors

. . . .

The wave of environmentalism washing over the U.S. petroleum industry will affect long-term prospects for the gas processing industry more indirectly than directly, as compared with other industry segments.

Because gas processing operations generally are seen as more benign and less visible to the public, their simpler scope of process plant operations won't have to deal with many of the complexities of their counterparts in other sectors of the petroleum industry.

However, natural gas liquids markets will undergo significant changes as a result of environmental initiatives in the 1990s.

As discussed previously, NGLs will be in increased supply because of the push to develop more clean-burning natural gas. And NGLs will be in greater demand as motor and heating fuels, spurring their premium value as petrochemical feedstocks.

NGLs will benefit from new legislation mandating cleaner burning fuels in some regards, but not in others. Although gas processors have feared that new limits on volatility will force a glut of butanes (the principal source of that volatility), there is reason to believe that increased LPG demand as a motor fuel and heating fuel will offset that.

The impetus of environmental initiatives' effects on NGL markets is turning around a previously pessimistic view of the industry's outlook for the 1990s.

The Gas Processors Association (GPA) in late 1989 noted that NGL prices had bottomed out, and predicted they would rise by 1991. There was a hint of that in winter 1990, when a sudden, severe, and extended cold snap in the U.S. Northeast saw propane prices jump the farthest in recent memory.

Its own demand growth aside, NGL value will rise as it tracks an increasing value placed on natural gas. GPA expects processors' margins will lag increases in gas prices by six months to a year, as the so-called gas bubble effectively dissipates in 1990.

The outlook for petrochemicals is mixed, however, as the likelihood of overbuilt capacity in that industry by the mid-to-late-1990s tends to depress demand for NGL feedstocks.

Still, if there is any segment of the petroleum industry that has reason for optimism under the emerging new environmental consciousness—without the regulatory headaches still facing the gas industry—it would have to be the producers of NGLs.

12

THE PUBLIC, MEDIA, AND CRISIS MANAGEMENT

The petroleum industry is arguably the world's most important. The Automobile Age that propelled the world's growth in the 20th century would not have been possible without the inexpensive, widely available range of products that fueled it.

In its relatively short history, the petroleum industry has almost always managed to deliver its vital commodities to the consumer safely, efficiently, and inexpensively.

A similar case can be made for how the industry conducts itself in day-to-day business. In the years that I have dealt with people in the petroleum industry, I have come to regard the majority of them as professionals of generally high levels of competence.

One can speculate endlessly over why this might be so—the fraternal clubbi-ness dating back to the Seven Sisters, the adventuresome risk-taking of the wildcatter, the highly competitive nature of the industry today—but it is a truism that the industry usually attracts some of the best and brightest in its primary disciplines as well as outside them. The public relations and media professionals within the industry generally are no exception to this truism.

Industry's Image

. . . .

If that is the case, then why does the petroleum industry have such a poor image with the public and the media?

I call it the "J. R. Ewing" syndrome. Like its namesake, the unscrupulous character on television's "Dallas," oil men routinely seem to be regarded by

287

the public at large as ruthless, greedy, and indifferent to social concerns. Despite its contemporary connotation, this phenomenon is nothing new. It certainly predates Exxon Valdez, or Santa Barbara, or even the Seven Sisters era.

The roots of the petroleum industry's poor image are in its monopolistic beginnings. Remember from high school history texts the political cartoons depicting bloated, sneering, greedy plutocrats dripping with oil? There is a direct line from those old John D. Rockefeller/Standard Trust days to the scathing editorial cartoons that appeared after the Exxon Valdez, only the latter came with a "raped Mother Nature" twist.

After the Teapot Dome scandal scarred the Harding presidency, muckraking novelist Upton Sinclair wrote *Oil!*, a popular novel wherein the son of a wealthy oil tycoon comes to learn that politicians and oilmen are an unscrupulous lot; thus, he turns to socialism.

That sort of theme runs through much of American media this century. Even the raffish charm of wildcatter Clark Gable in "Boom Town" was overcome by the character's ruthlessness.

Moviegoers of that era also got to see the industry's exploitation of Native Americans, the oil leaseholding Osage in Oklahoma, in the James Stewart starrer "The FBI Story."

The years after World War II proved a respite for industry's negative image, if only because it rode the coattails of the general go-go pro-business spirit of the era. The good, grey eminence of the Seven Sisters presided over an era of boom and bust and generally cheap energy prices, and the industry was personified by the Humble tiger and the singing service station attendants behind Uncle Miltie.

The popular media have always seized upon the industry's peccadilloes, as it has with many individuals and industries. But the petroleum industry has been notably effective at picking the worst possible times to hand the media just that opportunity. Recall the emerging environmental consciousness of 1968 and its resurgence in 1988 just before the major spills. That has continued today. For example, the Mega Borg tanker exploded in the Gulf of Mexico, threatening a 1 million bbl oil spill for a time just when Congress was trying to deal with spill legislation and President Bush was about to announce his decision on federal oil and gas lease sales off California and Florida.

Some days, it seems as if the petroleum industry can do no right. Exxon must be feeling particularly set upon these days, although to a significant degree that has been self-inflicted. A rash of spills in 1990 in New York Harbor led the company to suspend marine operations at its Bayway marine terminal at Linden, New Jersey, and its Bayonne, New Jersey, refinery and set up a task force to study and make recommendations on the situation. Exxon in early June 1990 released the results of that study, resuming marine operations and announcing plans to spend $10 million to upgrade safety and environmental performance at the facilities. A few days later, there was another sizable spill in the harbor—not related to Exxon, but you can be sure the linkage was there in the public's mind.

About that same time, Exxon released a cursory report by three leading British scientists with expertise on the long-term effects of oil spills in cold marine environments, who pronounced Prince William Sound recovering from the Exxon Valdez spill. At that same time, singer John Denver and National Wilderness Society President Jay Hair were holding up oily rocks to TV cameras in the sound *again,* with Denver angrily claiming that he does not trust anything Exxon says.

The industry's image suffers when it is held up in the media as a fount of greed and unscrupulousness. It suffers even more when it is depicted as a despoiler of the earth, because a threat to the environment brings the problem a little closer to home. But when does industry's image suffer the most, even to the point of overriding the other two images?

When it hits John Q. Public in the pocketbook. There is a sizable body of evidence in industry surveys that indicate the public's ire is manageable when it is perceived—rightly or wrongly—as corrupt or environmentally negligent when those issues do not affect them economically. In a sense this may seem cynical, but it is really just a common sense reading of how people generally react to issues of the day most strongly when they are directly affected economically.

Opinion Research Corp. (ORC), Princeton, New Jersey, has tracked public attitudes toward the petroleum industry from 1965 to 1989. It is clear that the most significant increases in an unfavorable view of the industry and parallel drops in a favorable view of the industry occurred after major spikes in the price of crude oil and then gasoline (Fig. 12–1). Although the industry's skid in public favorability started with the environmental boom, it gained its greatest momentum after the price shocks of 1971–73 and 1979. Conversely, the highest favorable ratings since the Arab oil embargo have come in the wake of the oil price slide of the mid-1980s.

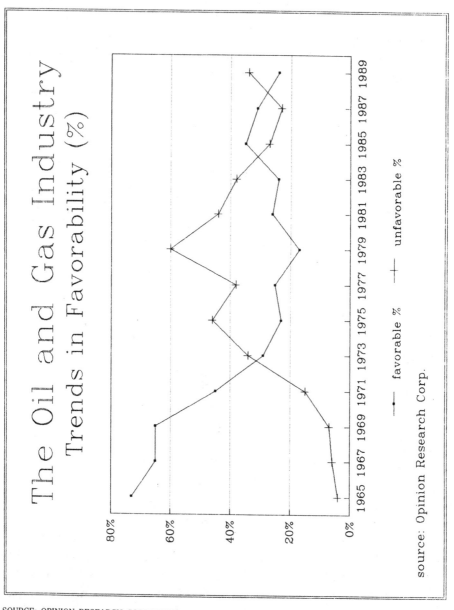

The Oil and Gas Industry
Trends in Favorability (%)

source: Opinion Research Corp.

favorable %   + unfavorable %

SOURCE: OPINION RESEARCH CORPORATION; *CORPORATE REPUTATIONS TODAY* research program; 1965 through 1989

NOTE: Data are based on samples of between 1,000 and 7,000 adults per year.

ORC's research indicates industry has suffered substantially since the Exxon Valdez spill and the reflowering of environmentalism. About three out of five people surveyed in March 1990 viewed the operations of the chemical and oil/gasoline industries as very harmful to the environment, faring the worst of the 11 industries studied. (Note the relationship of the third-place finisher to the petroleum industry.) (Fig. 12–2.)

ORC's surveys are based on telephone interviews with a random, scientifically selected nationwide sample of 1,046 adults split almost evenly between men and women.

Recall the outrage in Alaska shortly after the Exxon Valdez spill? Residents there were wearing black armbands, telling reporters they felt as if there had been a death in the family. They spoke of being in shock, mourning a loved one. Residents of fishing communities were especially angry, fearful of a loss of their livelihood. At one point, the radical environmental group Greenpeace attempted to spur the fishermen in the Prince William Sound area to blockade the Valdez Narrows with their fishing boats to keep oil tanker traffic from

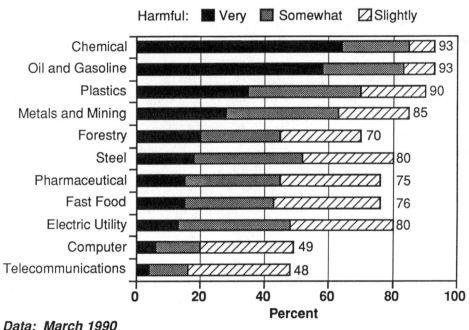

Data: *March 1990*
Source: *Opinion Research Corp.*

Figure 12–2 Environmental harmfulness of industries' operations.

the port. Despite their anger and grief, the fishermen turned down the Greenpeace request. Such a blockade would effectively shut down almost 25% of the nation's oil supplies, eliminate all but a fraction of the revenue stream that accounts for almost 90% of Alaska's revenues, and, by pinching North Slope producers' cash flow, perhaps complicate the fishermen's efforts to be reimbursed for losses incurred by the spill.

No one had more of a right to be angry over the Exxon Valdez spill than Alaskans. That anger was certainly in evidence for weeks and months to come. A poll conducted by the Alaska Oil & Gas Association in October 1988, five months before the spill, indicated that 69% of Alaskans had a positive view of the petroleum industry, versus 6% with a negative view. Shortly after the spill, another AOGA survey showed the positive rating fell to 50% and the negative rating jumped to 25%. By September 1989, another survey showed the approval rating had rebounded to 66.5% while the disapproval rating slipped to 21.5%. Bear in mind, the last survey occurred as controversy erupted anew over Exxon's winter plans for cleanup activity in Prince William Sound.

Ironically, one possible factor in all of this is that the spill actually boosted Alaskan revenues for the year because of the short-lived spike in oil prices that followed. It would be too superficial an assessment to conclude that Alaskans' anger subsided solely because the spill was not damaging them economically. But I suspect that was a factor.

It was not a case of short memories, because the spill dominated much of the national news and certainly the state news as controversy over the cleanup bubbled all summer long. At year-end, UPI American editors placed the spill as the fourth top story of the year, following only the California earthquake, reform in eastern Europe, and Tienanmen Square. The spill story beat out Hurricane Hugo, the abortion and flag-burning debates, Jim Bakker and Oliver North convictions, and the Sioux City, Iowa, plane crash.

Time and again, Americans' attitudes toward the oil and gas industry have been linked to the price of energy. Much of the outrage following the Exxon Valdez spill was inflamed further by the spike in gasoline prices. Even more outrage was stirred by the revelation that Exxon would be able to deduct the costs of cleanup from its taxes, leading to legislation being introduced to prevent that sort of thing, even though this has been an acceptable practice for many years.

What of the flip side, the more favorable view the industry enjoyed since the oil price collapse of 1986? Certainly, some of that can be attributed to the sympathy factor for the small independents and unemployed oil field workers and the economic devastation to oil state communities. But where was the gush of sympathy that would have produced an "Oil Aid" concert? Perhaps it was buried by the memory of almost $2/gal. gasoline and the bumper sticker "Let the Bastards Freeze in the Dark."

No doubt those were considerations. But I think it was simpler than that. The average consumer was simply delighted that the cost of a vital commodity was going down after a decade of often sharp increases; even more so at the sight of the price of *any* commodity going down in these inflationary times. What if the OPEC nations had declared an all-out production war in 1989? What if the price of crude oil had plunged to $10/bbl again? I don't think it is too much of a stretch to say that the anger over the Exxon Valdez would have been mitigated somewhat by the sight of 50¢/gal. gasoline again.

Changing Industry's Image

• •

I suspect the industry has not delved deeply enough into the question of whether, in fact, the drastic cuts stemming from the price collapse contributed to the 1988–90 string of accidents that so marred its image. I think it is imperative that industry put together a task force to study just that question realistically and unflinchingly and make appropriate recommendations.

It should not fear the critic who accuses the industry of putting profits before safety and environmental concern. Don't underestimate the American public's understanding of how a corporate economy works; each consumer has his or her own economic agenda as well. They recognize the exposure to added risk through belt-tightening, for example, in buying a thrifty subcompact instead of a gas guzzler—which in turn increases one's chance of dying in a car crash. In any event, the industry must answer that question, because it already is being forced to spend more money on environmental concerns. Finding the answer to that question will help direct it in finding solutions to recurring environmental and safety problems instead of just throwing money at them.

The petroleum industry must reeducate itself to not conduct its operations as if in a vacuum—a holdover from the chummy Seven Sisters era. This attitude tends to be smug and patronizing and more than a little arrogant by assuming

293

that "We can do our job just fine, if we're just left alone, thank you, and you need not bother about the details."

Certainly, the industry is doing a much better job of staying away from that sort of attitude, but there remains too much of its residue. Companies ought to take a close look at themselves and decide whether they still have traces of that attitude lingering about. In my encounters with Exxon's competitors, notably in Alaska, I detected a growing consensus that such an attitude at Exxon was what contributed to an environmental crisis turning into a public relations disaster.

I think the industry can learn much from Exxon on how not to handle a public relations crisis. Now it must learn how to do it right.

Dealing with the Media

· ·

Management at some petroleum companies seem to regard media relations as something of a necessary evil. They would just as soon not have to do it, but sometimes, grudgingly, they go through the motions: a press release after the fact, a designated public relations person instructed to hew to a lawyer-approved script to field inquiries, a "No comment," or vague fluff that really answers nothing.

Other companies employ big P.R. staffs whose designated function seems to be: Please submit your questions in writing and we'll get back to you on that (which is usually in the form of useless generic answers—and all too often after deadline).

Then there are those few antediluvian outfits who have no media relations whatsoever and will fire any employee who communicates with the press.

If some petroleum companies put a fraction of the effort into their media relations as they did their dog-and-pony shows for financial relations, they would be better off with both communities.

There are companies that are exemplars at public relations. ARCO, BP, Chevron, to name a few, are those that have impressed me not just in the competence of their P.R. staffs, but also in the company philosophies regarding media relations.

The companies that do well at media and public relations are those that require their executives in management and operations to be accessible to the press, train them accordingly, and are straightforward and responsive. Too many times, companies employ P.R. professionals as buffers against the outside world, and they simply earn a reputation for stonewalling that serves them ill in terms of a crisis.

Don't think that this sort of attitude does not come with a price tag. In the early days following the Exxon Valdez spill, when Exxon's poor handling of media inquiries contributed to an overall negative image of how it was handling the crisis, that company's stock value fell by more than $1 billion. Bad P.R. may not have been the sole cause of that loss, but it certainly was a contributing factor.

The company learned something of a lesson later. When it prepared for shutting down intensive cleanup operations during the winter, it placed a seasoned P.R. professional, familiar with the Alaska scene at Valdez, to coordinate a media tour of cleanup operations. The upshot was that instead of the newsmagazine doomsday covers that featured the spill in the spring, we saw a newsmagazine cover that referred to "the disaster that wasn't."

Much of industry's difficulties in dealing with the media are not entirely of its own making. Many print and especially electronic media are not equipped to provide the kind of technical training reporters need to cover some of the crisis stories that arise within the petroleum industry. Often, media outlets that have someone assigned to an energy beat generally just plug in reporters from some other business beat to cover energy just as they would have covered financial institutions or aerospace. When large metropolitan media outlets outside the oil patch states (California a notable exception) cover something like offshore lease sales or tanker spills, the coverage is by an environment beat reporter (certainly valid, if a business side reporter is also assigned to the same story to cover industry's angle on the story).

Consequently, petroleum industry news coverage by the daily media often tends to be superficial and often inaccurate. When oil companies are reluctant to be fully accessible or take the pains to be patient in telling their side of the story, it is only natural for a reporter to turn to another source who can tell readers just what the implications of an event are for the average consumer. The politicians with an ax to grind or career to promote and the environmental and consumer group lobbies with an agenda to promote or subscription rolls to fill are only too happy to oblige by filling that void.

It becomes a vicious circle. With this situation resulting in poor reporting, petroleum executives dealing with the next newsmaking situation come to the press with a chip on their shoulder. They've been burned by superficial, sensationalistic reporting by hostile or ignorant reporters, so they put up another stone wall—not realizing that their stone wall in the past contributed to that kind of reporting. Whatever goes around, comes around.

It is time to break the circle. Every petroleum company has an obligation to help the daily media achieve a better understanding of how that company and the industry works, how it goes about its daily business and what that means for the average consumer. It should also duplicate that effort at the community level as well. These companies should not wait for a crisis to occur to implement such programs. It should not be necessary for companies to wait for community leaders to bang on the doors of their executive suites seeking answers.

There is absolutely no priority more important in the 1990s than the petroleum industry's obligation to communicate its role on environmental issues specifically and generally to the public and the media.

Right now, it is not an exaggeration to say that the oil and gas industry has a fierce uphill struggle in this regard. If stronger initiatives are not taken soon, it will be a losing struggle.

A Few Recommendations

· ·

The first thing petroleum industry management must do in communicating a company's views on environmental issues of the 1990s is to ask itself: How best do we get our message across?

Is it enough simply to leave that assignment to the public relations department? Here, many companies automatically have a roadblock to overcome. Many of the P.R. staffs at oil companies are already overburdened and understaffed, a result of the industry doldrums of the latter 1980s.

Another problem is management philosophy. Too many companies go to the trouble of hiring highly qualified, talented, experienced P.R. professionals and then just do not let them do their jobs. P.R. is this distasteful chore that senior management would rather not do—the corporate analog of janitorial work—so it hires specialists to take care of the "problem." The real problem, however, is that a P.R. professional is doing his or her job best when it links a manager in operations or vice-president with a reporter and helps facilitate that communication.

Just as companies are restructuring to accommodate environmental concerns in operations, they should restructure at the public relations level by creating an environmental affairs (EA) office. Staff in an EA office should not have their efforts diluted by other duties that corporate, media, or financial relations staffers could handle as easily. This office should be the contact for all media on environmental issues and breaking stories. It should be this office's responsibility to maintain a working awareness of what the chain of command is in a crisis, to know which technical people and which executives would be the appropriate people to field detailed media inquiries, and how it should implement the company's messages and underlying philosophy on environmental issues.

An especially important measure is to see to it that the EA officer has not just access but regular input to the highest levels of management and the company's board of directors. The kinds of decisions oil and gas companies will be making in the 1990s on environmental issues will be make-or-break ones, and that should be reflected in corporate philosophies. A good approach would be to designate a special environmental communications committee consisting of board members, senior management, and EA staff.

The words "no comment" should be stricken from the vocabulary of anyone dealing with the media. If a company spokesman cannot answer a question because he does not know, then he should say so, and offer to find the answer and then get back to the reporter. All too often, though, "I'll get back to you" carries the unspoken subtext of "after your deadline." It is important to be honest in that regard. If you can't be sure about answering the question truthfully and accurately within a set time, make that known as well.

With that in mind, a company should familiarize itself with the media it deals with. Get to know deadline constraints, need for photographs or other visual aids, what a reporter's strengths and shortcomings are on environmental and general industry issues.

A decade or so of reporting and editing news about the petroleum industry for the leading trade journal allows my sources to be more presumptive about my knowledge of the industry and who my readers are than they could be for a newspaper or television reporter who might not ever have covered an oil story. Even an experienced energy reporter in the daily media cannot be expected to have background to deal with some of the issues that come up without someone with expertise in the industry to provide guidance. Even the most seasoned among the oil trade press still seek guidance and understanding. It is a tough and complicated business.

Accommodate different media needs differently, and don't play favorites. Most print reporters recognize that television and radio reporters have more urgent deadlines and less time in which to tell a story. TV reporters in particular also have special demands because of their equipment and the need for certain controlled conditions. But calling a press conference that allows for a few quick sound bites for the cameras and then walking away from the more detailed questions after the cameras have packed up and left is the quickest way to invite a print reporter's hostility. That in itself is often seen as a sort of stonewalling. Press kits and backgrounders can be a godsend in such situations, for all types of media. But they should not be seen as panaceas. The personal touch is still required.

How you communicate is of critical importance as well. It is a good idea to search for the key management people who have the authority to speak on behalf of the company in most matters and who are adept before a camera or a roomful of eager reporters. Training is essential, especially in programs that simulate press encounters, whether one-on-one or in crowded press conferences during a crisis. Workshops and training seminars should be built around such simulations.

Such programs are especially helpful in rooting out some of the most common problems reporters encounter in reporting on technical or otherwise complicated issues. Jargon is the death of clear communication. It is not patronizing to express technical issues in everyday language. Your average newspaper reader or television viewer is not an engineer. They can tell, however, when someone hides information behind a fog of technical terms.

Most importantly, take the initiative! If a reporter has been on a tour of your refinery and gone back to his office with photos, graphs, charts, background sheets, and conversations with the people likely to be his contact on a breaking story, isn't it logical that he will do a better job of reporting a story involving your refinery when a story does break?

Keep reporters abreast by issuing lists of contacts for possible stories and their areas of expertise. Call up reporters and ask them if they might be interested in a tour or an interview. Offer to help with your company's expertise on a general issues story, even on a story that may not involve your company or facility directly. Pitch story ideas. Some of the better stories I've developed for the *Oil & Gas Journal* over the years were first generated by public relations or even operating personnel within the companies.

Challenge of Crisis Management

. . . .

Is there a petroleum company around that still believes a crisis management organization is not essential for operating effectively in the 1990s?

Unfortunately, there are—too many. According to industrial communications specialist Alex Stanton, president of Dorf & Stanton Communications Inc., New York, only about half of the Fortune 1000 companies in the United States had a formal crisis communications plan in early 1989. The situation is even more dismal among small companies, he said.

Stanton maintains that most companies are totally unprepared to cope with a full-fledged corporate calamity, thereby raising the potential for a crisis to turn into a disaster when all the company can do is improvise.

Stanton says the first step a company should take is to designate a crisis management team, ideally a small group of quick decision-makers who represent all key functions. This team should develop a comprehensive crisis plan and playbook in advance and "live it, minute by minute, hour by hour," he says.

"These should be people who operate well under pressure and are fully familiar with the crisis plan and playbook. Crisis team members should be accessible to one another on a 24-hour basis via a wallet-sized telephone list of beepers."

Stanton contends the first priority of the crisis team in a crisis is to define the scope and severity of the crisis. There should be no barriers to information for the team.

This is best accomplished, says Stanton, when the crisis team meets to review the situation in half-hour or hour sessions during the crisis.

It also is critical to identify key target audiences affected by the crisis and then decide who needs to know what. What Dorf & Stanton recommends is a cascading notification system in which the most important audiences hear first and the least important last. This system should be spelled out in detail and rely on key people within an organization for passing information along. The crisis team itself should not be responsible for all the communications while it also is managing the crisis, Stanton says. At the same time, a system should be in place for monitoring reactions to the crisis and new developments.

In communicating on a crisis, companies must be quick, open, and honest. Stanton also counsels against speculation of unverified information or placing blame. The emphasis instead should be on solutions to the problem at hand.

All too often, a company's employees are the last to know what is happening during a crisis.

"It also is important to recognize that every employee who is properly informed and armed with the correct messages can be a significant soldier in your communications army," Stanton says.

"So while you may expose middle management and rank and file employees to different levels of information, all must be told promptly and told the same story. Lack of a party line can be fatal to a crisis management effort."

Stanton also cautions against relying too heavily on legal counsel during the crisis. While a legal adviser may be an important member of a crisis team and consulted throughout a crisis, management should not be afraid to move against legal counsel when warranted.

"Ultimately, management has the responsibility to preserve the reputation of the company and its special relationship with customers.

"Attorneys often will seek to limit liability by releasing as little information as possible and moving more slowly than the situation demands. That's not always the right decision for the company or product involved."

Stanton also advises the need to keep a log of events during the crisis, so a company can tell its side of a story later. It is just as important, he contends, to issue a final statement after the crisis that specifies what steps are being taken to avoid a repeat of the crisis.

As Stanton emphasizes, the three key words in effectively managing any crisis are communicate, communicate, communicate!

Mobil's Challenge

· ·

A good example of how to respond to a P.R. crisis and the effectiveness and wisdom of using outside P.R. consultants is the campaign Mobil launched in early 1990 to combat Measure A in Torrance, California, mentioned in the previous chapter.

Facing a negative image and an uphill scrap in defeating the measure after a series of accidents at the refinery, Mobil hired one of the top public relations firms, Braun & Co.

Measure A—limiting hydrofluoric acid storage to an impossibly low level by any business in the city—received national attention with parallels to Bhopal raised. Early polls indicated the measure would pass 3 to 1.

Braun talked to the Torrance refinery manager and conducted in-depth ascertainment interviews with about 50 community leaders. These one-on-one interviews identify the client but don't attribute the comments solicited to identify problems and possible solutions. Braun uses that as the basis for a P.R. campaign.

Mobil created a risk management and prevention program for hydrofluoric acid use at Torrance, one of the few such programs in the state. It took out newspaper ads profiling employees and detailing safety programs at the refinery. At the same time, it set up a community advisory panel.

Mobil then started up a bimonthly publication and produced an audiovisual program and brochures, all providing detailed information about the refinery. The company also invited more than 2,000 people to tour the refinery.

Just before the election, Mobil's campaign kicked into high gear. It told voters what the measure really entails: an increased volume of hazardous chemicals used at and trucked into the plant, plus an $8-million lawsuit the company planned to file to overturn the measure. Mobil employees held coffee hours in their homes, and the company bought more advertising, direct mailings, and a 30-minute local cable television program.

Measure A lost by a 3-to-1 margin. Braun Executive Vice-Pres. Doug Jeffe contends that although city leaders were concerned about the costs of a lawsuit, "if we had not successfully addressed the safety issue, we wouldn't have won."

The challenge Mobil and other Southern California refiners face now on the hydrofluoric acid issue is expanding that approach throughout Southern California to combat the Scaqmd ban.

BP's Paradigm

I can think of no better way to illustrate the importance of effectively managing an environmental crisis and at the same time keeping it from turning into a

P.R. nightmare than to offer what I think is a paradigm established by BP America in tackling the February 7, 1990, American Trader spill off Huntington Beach, California.

BP's crisis response was outlined for me by Chuck Webster, the company's manager of crisis management, as "how we managed the crisis instead of reacting to it:"

> We have two other full-time professionals on staff here. Our responsibility is to develop plans in the event of a crisis to go beyond what a local facility or business might be capable of effectively managing.
>
> The concept is that businesses within a company like BP America are designed to be lean and mean. We recognize that when you deal with the products that we do, whether crude oil, petroleum products, or even animal food, you face the potential for getting involved in a crisis that requires many more resources than you have in your business or facility.
>
> That was the starting point for developing what we refer to as a sort of "National Guard" of people with specific functions and skills to be called upon should you face a crisis that goes beyond what would normally be expected of that business or facility.

BP was training for a crisis such as an oil spill in the weeks before the spill. It had put its new crisis management team through several drills, including a major crisis drill in December 1989 that simulated an alky-feed butane leak from a pipeline that involved a flashback and secondary explosion with civilian casualties.

In January, Webster's team went through incident command system (ICS) training in Alaska involving an Alyeska pipeline drill in Valdez. ICS was adapted from an emergency response management approach originated by California firefighters.

"ICS allows you to go, in fire parlance, from a one company response to a multicommunity response while maintaining an organizational structure that allows you to manage a situation," Webster said. "It has an incident commander at the top; an operations section; a planning section; a logistics section; a financial section; and subsets below that to allow you to grow as much as you need to grow, and cover everything from air operations to who's going to make sure there is food in the command center to lodging to people who can work skimmers—every phase of the operation."

It helped that the cities most immediately affected by the spill also happened to use crisis management systems based on ICS, helping effective communications.

"What is critically important," said Webster, "is to have local people from the mayor to the police chief to the fire chief to city leaders understand what you are doing to protect the environment. There have been times in the past that that aspect of communications liaison has not worked effectively. In this case, it worked very effectively."

BP sought input from the communities on priorities for beach cleaning, thereby advancing the opening of some beaches.

"This was appreciated by them and it allowed us to be more sensitive to local community interests."

BP continued the emphasis on local input. Said Webster:

> For instance, in addressing the major beach contamination, there were many options to get oil off the beaches, from manual labor to front loaders. Before making a decision, we talked not only to state and federal officials, we went all the way down to lifeguards who lived on the beach and asked them to fill our heads with as much information about the makeup of the beaches and their history before we made any decisions. One of the key things that we discovered is that there is an erosion history to those beaches. So one of the things we decided to do is go for more manpower, to stay away from heavy machinery to reduce the removal of sand. It caused us to choose technologies such as pom-poms and absorbent sausages that got the oil up without removing a lot of the sand in the process. It was absolutely the right kind of decision for that environment, but one that would have been much more difficult to arrive at had we not taken the organizational approach that we did.

Timing was critical in the response. The BP crisis team was on a plane headed west from Cleveland within two hours of notification, which came 30 minutes after the spill occurred.

"It is BP's policy that whenever there is oil in the water, we send a team with the view that, until we know for sure what caused the oil to be in the water, we will assume there is a hull fracture or something of an equally serious nature," Webster said.

"As a result, we have sent teams out on spills of less than one barrel, with the view that we would much rather come back for dinner than to find out that we should have been there sooner."

Before dawn the following morning, top BP officials were being briefed by the Coast Guard and talking to reporters—including an appearance on national television's "Today" show. BP began calling in more than 100 professionals that comprise its crisis response team. Webster said:

> We also made decisions early on that we were going to be absolutely available to the media, that we were not going to dodge questions, and that we were not going to try to put any 'spin' on the story. And that meant that when things didn't go well, we put them out with the same gusto that we tried to report successes.
>
> There was even a point where there was a system set up with state and federal officials to do surveys of the beaches to determine which areas were clean and which areas required additional work. We reached a point where we were finding buried oil in areas where the survey teams declared the beaches clean. So we put out a news release that we would like to see a more comprehensive survey. We weren't satisfied that all the oil was being found. It helped to develop credibility with those who were covering the story. And it just served the purpose of presenting ourselves as people who had a good measure of pride and wanted to see the job well done. I don't think that we made any promises that we couldn't keep.

To sum up the philosophy behind BP's crisis response, Webster said, "Information is your biggest ally and your biggest enemy. The more people understand about what you are doing and why you are doing it, the better shape you're in.

"We found in our team just a bunch of very, very skilled people who were within an organizational structure, given a task, and allowed to do it—and the results were very gratifying. There was a lot of pride of ownership in this job. We won some respect from local communities."

BP found a gauge of that respect when it returned a month later to Huntington Beach to throw a dinner for about 100 locals to thank them and tell them it will carry on the process of "lessons learned."

"I couldn't believe it," Webster recounted. "Here we were, in *Southern California,* no less, and here were all these people coming up to us and thanking us for what a wonderful job we did."

The upshot? As indicated in an earlier chapter, polls of Californians shortly after the spill suggested that a majority did not feel especially concerned about an immediate, pressing environmental crisis affecting California. Most today could not even name the American Trader.

The tanker that ran aground at Bligh Reef in Prince William Sound, however, has had to undergo a name change—Exxon Mediterranean, to show its new venue. And the name of the Exxon Valdez, like the Santa Barbara oil spill, will resonate forever to industry's chagrin.

13

BATTLE PLANS

What is the goal the petroleum industry should strive for in the Decade of the Environment?

To put the environmental lobby out of business.

There is no greater imperative. This is not some sinister call to arms for corporate skulduggery or guerrilla tactics to destroy groups bent on bettering the world for the sake of preserving profits.

If the petroleum industry is to survive, it must render the environmental lobby superfluous, an anachronism. It must seek to make the environmental lobby and its political proxies totally irrelevant.

That is not to say that environmentalism should be made irrelevant. After a second rapturous embrace in as many decades, it is time to imprint the spirit of environmentalism permanently into the fabric of American culture.

Real change is beginning to happen in this regard, not merely a chorus of lip service from America's boardrooms. The environmental movement's momentum likely will be slowed in the 1990s after the megahype of 1988–90 and after economic and other geopolitical issues displace it on the world's center stage.

But it will not recede into the shadows as it did during the Reagan years. Just as the legacy of the first Earth Day left environmentalism a strong part of the American ethos, this current revival will enshrine its permanence. If nothing else, there are very few cultural values left to Americans that are no longer in doubt. Outside of the American family, there is perhaps no other cultural value that is still as safe a haven from skepticism.

The question becomes one of degree. It is time for the American public to learn that there is a true environmentalism that can coexist with profit, that it is simply good business to be environmentally conscientious.

It is also time to recognize that American consumers must be educated about the need to integrate science and economics into the spirit of environmen-

305

talism—to make environmentalism more of a science and more of a business and less of a religion and political cudgel.

Industry must move beyond appeasement of the ecopoliticos. If it truly embraces the environmental spirit and conducts its day-to-day operations accordingly, the need for that appeasement will end. Consequently, the need for the existence of the environmental lobby will effectively come to an end.

Do not hesitate to confront and do battle with the Neodruids. They do not possess the holy writ on what determines an environmental value. In fact, some of what they advocate may even be more harmful to the environment than the status quo. It thus becomes a moral imperative to ascertain the truth behind actions aimed at preserving the environment.

There is another moral imperative in dealing with American environmental concerns, one that applies also to the rest of the industrialized world. The industrialized nations, led by the United States, have exploited much of their own resources and bruised the environment in their climb to affluence and a high standard of living. There remains a consensus among the environmentalists and the governments of those nations that environmental solutions must truly be global to be effective.

Is it not arrogant to presume that developing countries must somehow leapfrog past industrialization to embrace the costly "limits to growth" environmental mentality, to impose upon them environmental standards we have only just now devised? Should we "export" our pollution to them as our own polluting industries are legislated out of existence and we import the products of their industries that rise to take their place?

The leaders of the developing nations are justified in questioning this bit of ethical imperialism. Who is to say that it is simply not just another effort to keep their economies subjugated to a permanent market ghetto? Nations that are drowning in debt and cannot afford to feed their people cannot afford the luxury of locking up resources whose development might bring them into the 20th century. Halting the destruction of rain forests may seem a noble pursuit to comfortable Americans, but it could mean starvation, poverty, disease, or death to the farmers, ranchers, and villagers without electricity or running water who might need that land.

It is all well and good to talk about the need to put a lid on energy demand through conservation, but energy demand is going to rise no matter how extensively the United States implements feasible conservation measures, by most industry forecasts. The reason: World population will more than double to about 11 billion people by the year 2100. According to retired Shell Oil Co. Pres. John Bookout, fossil fuel production must increase by more than 50% during the next 40 years even with energy conservation measures in place. By about 2030, Bookout contends, the United States will have to turn to new energy sources, cut energy demand further, or find answers to environmental problems that otherwise would limit fossil fuel production.

That last option will be the petroleum industry's principle mission in the 1990s. Why? The industry has an obligation to continue expanding the wealth of its shareholders and contributing to the U.S. and world economic well-being.

According to a study by Prof. Dale W. Jorgenson and Dr. Peter J. Wilcoxen for Harvard University's Energy and Environmental Policy Center, early studies of the economic costs of environmental regulation overlooked the important role of reduced capital formation and thus underestimated the impact of regulation on long-term growth.[1]

Their study found that environmental regulations have cut the annual growth rate of U.S. GNP by 0.191% during 1973–85, several times the reduction in growth estimated in previous studies. The cost of environmental regulation was a long run reduction of 2.59% in the level of U.S. GNP, a figure greater than 10% of the share of total government purchases of goods and services in the national product and significantly larger than that reported in earlier studies on this subject.

Jorgenson and Wilcoxen found that the decline in economic growth stems from mandated investment in pollution control lowering long run capital stock, reducing long run consumption, and lowering the rate of capital accumulation, especially in the early years of the regulation.

They cite notably mobile vehicle emissions controls, which they contend account for 27% of the long run reduction in U.S. GNP. Vehicular emissions controls have slashed the long run output of the U.S. auto industry by 15%. Other industries similarly affected are oil refining, coal, metal mining, and pulp and paper, they found.

Perhaps we could stop the Japan-bashing and attacks on mismanagement or lack of worker productivity to consider this as another reason for America's inability to compete.

The industry also has a moral obligation to help the developing nations of this world raise their standards of living and join in the economic bounty the industrialized world enjoys through sound, benign development of their resources. (It will be fascinating to watch the growth of countries like Yemen and Papua New Guinea, suddenly emerging from destitute status to become significant exporters of crude oil in the early 1990s.)

In what is an exciting prospect for balancing human and ecological needs, a new conservation ethic may be emerging: the use of biospheres, which mix conservation and development. The idea is to set aside a parcel of environmentally valuable land in a rain forest as a wildlife preserve, surrounding it with a buffer zone designed for research, land restoration projects, and native subsistence use. That is then surrounded by another buffer zone of land for multiple use development, such as logging or agriculture. This way, scientists can find out just how much land is needed to maintain a viable ecosystem and still accommodate a certain amount of development. The pristinists won't like it, since their stance is non-negotiable, but a developing nation's leaders may not wish to kowtow to the trendy urges of celebrities in California when the starving masses in their own countries are near riot. Perhaps such biospheres could serve as examples for selective sanctuaries amid oil and gas exploration and development in the United States.

Finally—and what may someday prove the most important point—the industry, which still is guided to a large degree by men and women trained as petroleum *scientists,* is obliged to ensure adherence to sound science to ensure that so-called environmentally preferred solutions touted by lobbyists, politicians, and social engineers are not an environmental Trojan Horse.

The same denouncers of pesticides, which curb the presence of natural carcinogens in produce 100 times more carcinogenic than the pesticides, disregard the increased potential statistical risk. Who is to say what such a radical altering of the American diet might cause? Who is to say what the almost unfathomable workings of atmospheric photochemical reactions might be as we favor cutting certain air emissions? Some studies suggest cutting certain targeted man-made emissions, while disregarding emissions from natural sources, could actually worsen the greenhouse effect.

An operative guideline here might be that which serves the medical community: First, do no harm.

Educating the Public

Every day, it seems, there is another poll showing how strong the degree of environmental concern is among Americans.

308

These polls tell us how much Americans are willing to sacrifice to reduce pollution. Overwhelmingly, Americans polled say they are willing to pay higher prices, do without, and suffer inconvenience to cut down on pollution.

Those same polls indicate, however, a divergence on pocketbook and energy security issues.

A survey by Cambridge Energy Associates Inc./Opinion Dynamics Corp., Cambridge, Massachusetts, taken in December 1989 of 1,250 adults, found that three out of four persons would opt for environmental improvement over economic growth if forced to choose between the two.[2] This view cuts across party, income, and education groups. Furthermore, those surveyed by CERA generally say that big business and government tend to favor the economy over the environment when choices have to be made. An especially telling point, however, is the view by a two-thirds majority that we can have economic growth and a cleaner environment at the same time.

CERA's poll also found, however, that there is widespread concern over the issue of energy dependence among Americans. About seven in 10 think a high priority for government should be making the United States less dependent on foreign oil. When people were asked whether drilling should be banned in the ANWR, 57% said yes, which CERA called a "surprisingly slim" majority in light of the Exxon Valdez spill only nine months earlier. In addition, 55% opposed restrictions on offshore drilling if it meant increased foreign oil imports.

People were also asked whether they feel more threatened by energy shortages or by the environmental problems caused by the use of oil, coal, and gas. Although environmental problems were seen to be the greater threat, the margin was only 4-to-3.

"These findings suggest that if energy problems were once again to dominate public perceptions as they did in the 1970s, then the environmental consensus would face strong competition and greater challenge," CERA said.

"Such developments as rising energy prices, electricity interruptions, a continuing sharp fall in domestic oil production and a variety of other issues could all affect the strength of the public's commitment to environmental improvement Cera concluded."

A leading indicator of this possible trend is how Washington is salivating over the prospect of raising taxes in the name of the environment. After Californians in June 1990 approved a 9¢ gasoline tax increase—in the birthplace of the U.S. tax revolt—Washington politicos scrambled to devise new environmental taxes, generally placing a tax on things like air emissions or carbon content or water discharges or nondegradable products. The concept that has them salivating is that such a tax is a "user fee" for polluting the environment.

However, industry recognizes that such environmental taxes simply penalize U.S.-based production and hinder U.S. competitiveness. Further, proponents of environmental taxes disregard surveys at the time that clearly showed Californians favored a closed-end tax like the one they approved because it specifically targeted the state's crumbling transportation infrastructure and was not aimed at pollution or conservation.

The petroleum industry must press a massive education effort to instruct Americans on the costs associated with environmental regulation and their resulting effects on U.S. energy dependence, as well as the repercussions through the economy.

The National Association of Manufacturers in April 1990 released a study that showed America has spent more than $1 trillion—two-thirds of that spending by manufacturers—on pollution abatement, control, and prevention since Earth Day 1970. Industry spending on environmental mitigation now totals about $49 billion/yr, NAM estimated. Total employment and GNP are about 50% larger than in 1970, yet five of six target pollutants have dropped substantially during that period.

Does that mean that environmental protection does not stifle economic growth? Recall the view that holds American business and government have not done enough for environmental protection. Could these numbers just as easily mean that without the balanced approach toward environmental regulation during those two decades, America might not have had that economic growth? Or that onerous environmental regulation might have spurred a deep, long-lasting recession in which American business could no longer afford any significant environmental mitigation and stay competitive? Industry should seek to quantify the answers to these questions and make sure the public comes to learn them.

There is no better place to start the educational process than in American schools. A poll in spring 1990 by the National Energy Education Development Project showed that most high school and middle school students could not understand or identify basic concepts about energy and the environment involving costs, efficiencies, and fuel uses and properties.

Most of those students polled believed that renewable fuels account for a much larger share of the energy mix than they do. Will the next generation grow up thinking that U.S. energy problems will be solved by solar and supercon ductors when economics and science dictate those are at least several generations away and that fossil fuels must be a key to bridge that gap?

The petroleum industry could well emulate the grassroots movement to educate youngsters on the perils of drug use with a comparable outreach program on energy and the environment.

Fighting Back

. . . .

It is time to get tough with the Neodruids.

Petroleum companies must begin to think of extremist environmentalist groups as competitors for the same resources: sometimes directly, as in access to lands, sometimes for political capital, sometimes for the hearts and minds of the American people.

There is likely to be a backlash of public opinion against the environmental lobby in the next few years, provided no more Exxon Valdez-type disasters occur to keep the public ire galvanized.

It could easily move beyond discomfort over the leading neodruids' holier-than-thou attitudes and elitist nature of these predominantly white, affluent, highly educated lobbyists and their not-too-subtle social and political agendas.

The first real sign of widespread public dissatisfaction with the environmental groups may well occur as this book goes to press in late 1990.

There could be confrontations in the Pacific Northwest, where an epic battle is playing out between loggers and environmentalists over the northern spotted owl. If plans to harvest old growth timber in the owl's habitat proceed, many environmentalists plan to disrupt the activities.

What could get the most publicity out of all of this is ecoterrorism. Groups like Earth First! have been linked to environmental terrorism ("monkeywrenching" after the book by their spiritual father Edward Abbey) such as driving metal spikes into trees to disable equipment. A logger in California was almost killed when his high-speed saw hit such a spike in 1988.

The campaign by environmentalists to halt logging may be mostly peaceful confrontations, but the trend is toward escalating violence. In June 1990, two members of Earth First! were injured when a bomb exploded in their car. Although the investigation was proceeding at presstime under the assumption that the two allegedly may have been transporting the bomb for purposes of environmental sabotage ("ecotage" as the "Earth Firsters" put it), the two Earth Firsters insisted they were the target of the bomb and of an FBI plot to discredit them.

Violence, confrontation, and bizarre symbolic acts by the radical fringe of the environmental movement will only serve to taint the entire movement, just as the peace and civil rights movements experienced in the 1960s and 1970s.

When that occurs, it would be entirely appropriate for the industry to raise some questions about the environmental groups. Just exactly who are they? Who are their leaders, and what are their connections in government? Why do so many of them seem to wind up in government jobs related to environmental regulation of the areas they specialized in? Where does their funding come from? What are their connections with other political groups with perhaps less acceptable social and political agendas?

An especially intriguing question arises over funding. It would behoove industry to challenge the environmental groups to open their books and operations to public scrutiny. The industry won't be surprised to find the level of contributions from government, but it may prove an eye-opener to see how much the industry itself gives to the very organizations that so vehemently oppose it on critical issues. That hardly seems prudent. There are many worthy issues and groups, including other environmental groups that conduct constructive, nonpolitical work like the Nature Conservancy, that deserve that funding. Don't bite the hand that feeds you, the saying goes—so why keep feeding?

The petroleum industry should not hesitate to discredit these groups when they merit discrediting. All too often, petroleum companies are placed on the defensive, hemming and hawing in meek, flustered rebuttal.

Why not take the offensive? Ask, "Where are your facts, your research, your proof? Lay them out in detail. Submit them to an unbiased panel of experts. How does your group stand to benefit from your proposals?"

These are valid questions. The American public needs to be reminded that the environmental lobby has an agenda, subscription rolls to fill, books and magazines to sell, lawyers and administrators to keep employed, politicians and regulators to keep on cozy terms for future employment.

The California Coordinating Council, fighting the nightmarish "Big Green" initiative on the ballot in November 1990, lost no time in reminding the public that the measure also was intended to propel the careers of sponsors state Atty. Gen. John Van De Kamp (defeated in the Democratic gubernatorial primary) and Assemblyman Tom Hayden (formerly of Jane Fonda fame). Woodward & McDowell, the San Francisco political consultants the council hired to combat the initiative, conducted a poll of Californians on "Big Green" and recognized that those surveyed tended to recognize the politically self-serving aspects of the measure for its sponsors, notably pointing out Tom Hayden as angling for the job of Environmental Advocate (read "czar") that the measure would create.

That poll, incidentally, was especially instructive in proving the benefits of educating the public on the economic costs of environmental regulation. It showed that California voters reject the initiative when they learn of its contents. An initial survey on the measure had 76% of voters favoring the initiative and 16% opposed. That was followed by pro and con arguments and a second ballot. In the second ballot, 40% of those surveyed favored the measure and 56% opposed it—in California!

Are scare tactics fair game? Certainly, if they raise valid questions and are backed by science. A 1987 study was commissioned by Robert Danziger, General Partner of Sunlaw Cogeneration Partners, in response to a new rule proposed by California's Scaqmd requiring selective catalytic reduction technology to control nitrogen oxide emissions in stationary gas turbines. That study found the upper limit cancer risk of one mid-sized SCR unit to be 200-in-1 million—a factor 200 times the agency's own recommended standard for rejecting other technologies.

"In effect, SCR means trading a very small reduction in NOx for a potentially unacceptable increase in carcinogens," Danziger said.

Environmentalism can prove to be a Pandora's box in itself. Why should we hesitate to investigate the claims of nonscientists with obviously political agendas on issues that could affect the health and well-being of everyone on the planet? Why should we not demand the same sort of scrutiny on environmental as well as economic effects of their environmental proposals?

For example, the government's ban on the chemical EDB has removed the safest and most effective method of fighting the much more deadly carcinogen aflatoxin that is tainting the nation's grain crops.

Sometimes, it just takes common sense to recognize that an environmental "solution" sometimes creates a bigger environmental problem. For example, if the EPA declares used oil to be a hazardous waste, that almost certainly will result in fewer gasoline stations trying to recycle used oil because of the liability and costs associated with handling and storing hazardous waste. The result: More used oil will be dumped onto the ground or into sewers, resulting in more environmental damage than would have occurred with an exemption.

Perhaps it is time to press for legislation that mandates the same sort of scrutiny for a proposal to mandate alternate fuels or a study that claims the harmful effects of some product of facility as we do for a proposal to open ANWR or drill offshore.

Environmentalists who claimed Alar was a deadly poison risking the nation's children with cancer should have been required to submit their study to an independent scientific panel for review before orchestrating a media campaign that terrified the nation and nearly destroyed an industry. Under such legislation, if such claims are not supported by independent review by scientists, then those claims would be frivolous at best and blatant disregard of the public welfare akin to shouting "Fire!" in a crowded theater at worst.

Certainly they should be held accountable, just as any other business is held accountable for what it tries to tell the public under a panoply of laws and regulations. The Food and Drug Administration will hold up introduction of a new food product or medicine for years to ensure it will not harm the public even if there are clearly demonstrated benefits. Why should we expect less of claims against existing products that have a clear benefit? Why not investigate whether the anti-Alar forces have financial interests in growing and marketing organic produce, teeming with all those potent natural carcinogens?

If this sounds a bit far-fetched, consider the plight of the apple growers who lost millions. And who benefited? The kids that got junk food in their lunch boxes instead of nutritious, healthy treats because of ecoterrorism? How about the anti-Alar forces whose blaze of publicity netted them scores of subscribers and a torrent of donations?

Correspondingly, such legislation could provide a means for recourse against frivolous and destructive environmentalist claims. Products or facilities so maligned unjustly who could prove damages should proceed to try to collect them in a court of law or through an agency designated to arbitrate such claims. And just as celebrities who tout products or services on television are being held accountable for their endorsements, should they not be held accountable for the claims of environmentalist groups they front for?

If the environmental lobby wants the petroleum industry to adhere to the strictures of a code of environmental ethics such as the Valdez principles or the CERES code, why can't the industry in turn request environmental groups adhere to a code of ethics as well?

Let's call them the Alar principles. Environmentalists not willing to adhere to a code of ethics requiring a study of economic and human impacts of an environmental proposal, adherence to sound scientific principles reviewed by an independent body of experts, freedom from conflict of interest, and appropriate diligence and caution in making public such claims, should be branded as such and cut off from all government, foundation, and corporate donations.

Environmentalism is the new Lawyers' Full Employment Act. But that is a two-way street. If an environmental critic attacks without justification a company or a facility as a polluter, why should he be exempt from libel laws? Why cannot the employees of Exxon and Alyeska, so vilified by environmentalist publications, seek redress in the courts? Why can't campaigns of harassment and frivolous legal actions to halt or delay projects be met with lawsuits that fight back?

I personally don't care for this approach, because the last thing this country needs is more litigation—and it could backfire. But the environmental lobby has been using the courts all too effectively for years, and if it remains unwilling to compromise, this may be an effective last resort.

Far better for industry simply to take its case again and again to the American people and trust in their judgment. Mobil and BP did it successfully, as seen in the preceding chapter.

As former Interior Sec. Don Hodel noted in a 1987 talk before San Francisco Bay area recreation and tourism leaders, the energy crises of the 1970s wreaked havoc on tourism and recreation in the United States.

Citing catastrophic effects on tourism in the wake of the 1973–74 and 1979–80 energy crises, Hodel said, "When Congress chose to set oil allocation priorities, all sectors suffered, but recreation and tourism were particularly hard hit. Specific recreation activities were singled out as nonessential energy uses in excise tax proposals and government sponsored ads."

Hodel noted that California's large recreation and tourism industries experienced some of their worst years as a result of the 1970s energy crises, "when people couldn't or feared they couldn't get fuel."

The 1979–80 energy crisis alone cost 500,000 jobs in travel-related industries nationwide, Hodel noted.

"Thus a few oil rigs off the California coast are in fact the best insurance policy for the tourism industry," he said.

The petroleum industry should practice this kind of specific constituency politicking and take this kind of message to each group as energy issues specifically affect it.

The best approach is a grassroots approach, and the industry has a marvelous tool for that at its disposal: its own employees, past and present.

That is summarized in a speech on November 11, 1989, before the National Ocean Industries Association in San Francisco, by California Republican Rep. William Dannemeyer in his characteristically colorful way:

> I think it is time for the industry to go on the offensive and take their case to the people. The average citizen must be told that while he must suffer the consequences of energy dependence and oil shortages, the grass-eaters and daisy-pickers who created these political dilemmas will always make sure their cars start in the morning.

> These good neighbor "scams" are the same ones that lock up the California coastline to further development only *after* they have insured their own beachfront properties. The average citizen will understand this hypocrisy, and will most likely tell these eco-nuts to get a real job.

> Environmentalists have established a grassroots network across America whereby their political clout is much more profound than the political clout of the combined energy industries.

> The National Wildlife Federation alone has a membership of some 5 million persons. This figure is three times more than all retirees, employees, and stockholders of the nation's 12 leading petroleum refiners combined.

> The annual budget of the NWF is around $70 million, 62% of which goes to encouraging grassroots participation. From limited information offered by 12 leading oil companies, it looks as if they only spend a few million dollars on the grassroots. So we have a well-financed, active army of environmental grassroots driven by a deeply imbedded ideology versus an underutilized, sleeping army of consumers, employees, and oil businesses. I am convinced that this is a large part of our problem in our struggle for energy independence.

> The industry must fight fire with fire. It needs to tap shareholders, retirees, and annuitants to point out to local governments that ongoing bans on energy production affect jobs. I guarantee you that if members of the House were presented with pro-employment resolutions from their local governments, they would think twice about routinely voting with environmental extremists and against the creation of jobs for their largely blue collar districts.

> When this war is effectively waged, this nation will be able to get serious about energy dependence.

That is what the agenda should be: the industry keeping its own nose as clean as possible when it comes to protecting the environment and conducting a grassroots campaign to inform the public of the costs to the American economy, energy security, and competitiveness of onerous environmental regulation.

The industry should recognize that this is a jobs issue and thus conscript labor unions into the battle; that it is an individual economic issue and thus conscript those most disproportionately affected by high energy prices—the minorities, the elderly, the lower income groups; and that it is a consumer issue and thus conscript the ranks of the middle-class who fret about disposable versus cloth diapers and whether they'll be forced to spend thousands more for a methanol vehicle.

The industry should also hammer home the message that as the federal budget squeeze and trade deficit worsens, and more issues such as drugs, education, crime, and homelessness clamor for a piece of a shrinking pie, we can ill afford to throw money at environmental "solutions" that are counterproductive or not cost-effective.

Some Positive Solutions

. . . .

Think in new ways.

That surely must have driven Unocal Corp.'s innovative pollution program in Southern California, where air quality concerns threaten to drive out the petroleum industry.

Almost breathtaking in its obvious simplicity is Unocal's proposal to cut pollution in Los Angeles. Because 60% of smog comes from cars and trucks, and 30% of that from pre-1975 vehicles, Unocal earmarked $5 million to buy and scrap as many as 7,000 old cars in the Los Angeles area.

By offering $700 and a one-month bus pass apiece for pre-1971 vehicles, the program could make a small dent in vehicular emissions as a demonstration program, but a big one if the idea catches on and spreads. Simply finding a way to ensure that the recipients don't just replace their polluting cars with other old polluting cars would ensure that this part of the problem is largely eliminated. Tightening smog check tests and requiring them before resale of any vehicle could take care of that problem.

Unocal also will offer free tune-ups for pre-1975 cars in off years that smog checks are not required.

In addition, Unocal started a 76 Protech Patrol—aligned with its full service gasoline stations—wherein specially equipped vehicles roam California's freeways to assist at no charge motorists whose vehicles have broken down. Such incidents contribute fairly significantly to Los Angeles's smog problem because of the long-stalled, idling miles of traffic every day.

Not only does this program offer genuine environmental benefits and terrific marketing P.R., it can generate new business for its Protech stations. It is just about as perfect an approach to industry's dilemma in the Decade of the Environment as one can imagine.

Other positive solutions for the petroleum industry can embrace a cooperative approach with government, the public, and responsible environmental groups. Clean Sites, the neutral, nongovernmental organization formed by industry, environmentalists, and the public to clean up orphan waste dumps is one such example.

Cooperation is needed also on developing market incentives for reducing pollution instead of just pursuing regulatory solutions, whether they be pollution offsets and permits or hazardous material deposits.

Another concept worth noting is the creation of a national environmental trust fund, similar to the National Highway Trust Fund, to clean up orphan hazardous waste sites. This idea is being fosterd by American International Group, the biggest underwriter of commercial and industrial insurance in the United States. It could be financed, says AIG, by assessing a fee on commercial and industrial premiums that could raise as much as $40 billion to clean up priority Superfund sites. It suggests a national advisory board of private citizens, industry, and public officials charged with overseeing the program.

"Just think," AIG offered in a series of national ads, "A new way to finance Superfund's mission without the need for new taxes, a new government agency, or expensive and unproductive lawsuits."

Crossroads: An Energy/Environment Summit

I propose that all of the aforesaid be a prelude to a working summit on U.S. energy and environment issues.

On a smaller scale, such summits on specific issues have drawn praise and produced results, as can be seen in the cooperative efforts of the Institute for Resource Management (IRM), founded by actor/director Robert Redford.

IRM brings together leaders from industry, environmental groups, and government to discuss areas of disagreement on specific issues. IRM's goal is to seek balance in using and preserving America's natural resources. An IRM conference in 1985 led to a compromise plan that allowed drilling in the Bering Sea off Alaska.

If the petroleum industry must resort to hardball tactics to bring the environmental lobby down to earth, then it should follow that campaign with a proffered olive branch: an opportunity to invite all concerned parties to gather in common cause and at least establish the true parameters of the energy/environment agenda.

Why not promote a full-fledged summit that captures the attention of the nation and seeks to achieve consensus on what is needed to balance energy and environment concerns?

It could be held in some midsized, neutral city—not a Los Angeles or a Houston, where the sides would be too polarized.

It should invite leaders of the petroleum and other energy industries, environmental groups, government, regulatory agencies, consumer lobby and other public interest groups, economists, scientists, and academicians.

Put these individuals into conference rooms to conduct a bit of horse trading on the issues. Send them out into the wilderness on a retreat to get to know each other under casual circumstances.

Discover areas of commonality and try to emulate that in areas of disagreement. Agree to certain ground rules that hew to a rational, problem-solving approach.

Find out where the Valdez principles and the Alar principles intersect and perhaps hammer out a genuine accord that all can someday be signatories to.

We are working, ostensibly, toward a National Energy Strategy under DOE Secretary Watkins. I don't believe that can be accomplished without setting a National Environmental Strategy, an alternative to the bewildering jumble of piecemeal legislation and regulation that threatens to choke industry into a business coma.

Perhaps when Secretary Watkins' NES is formulated, it can form the basis for such a summit. The President, the Congress, the petroleum industry, responsible environmentalists, and the main of the American public must place this among the nation's highest priorities.

Energy and the environment: One without the other could be a grim legacy for generations to come.

We need not choose between the two. We can have both, and we cannot afford not to have both. It takes but the will to find the way.

References:

• •

1. Dale W. Jorgenson and Peter J. Wilcoxen, "Environmental Regulation and U.S. Economic Growth," Discussion Paper (E-89-14), Energy and Environmental Policy Center, John F. Kennedy School of Government, Harvard University, November 1989.
2. Cambridge Energy Research Associates/Opinion Dynamics Corp., *Energy and the Environment: The New Landscape of Public Opinion,* February 1990.

California State Lands Commission, 36, 38, 199
California Water Resources Control Board, 36, 38
Cambridge Energy Associates Inc., 133, 309
Cancer risk, 45, 60, 75, 278
Capacity creep, 237
Cape Resurrection, 14
Capital costs, 151
Capital outlays, 237
CARA. *See* Computer-aided risk assessment.
CARB. *See* California Air Resources Board.
Carbon dioxide, 100, 102, 142, 159, 205, 231
Carbon monoxide, 34, 84–85, 244–245, 257
Carbon tetrachloride, 34
Caribou, 174
Carnegie-Mellon University, 242
Carr, Donald Alan, 81
Carson, California, refinery, 251
Carson, Rachel, 33
Cascading notification system, 299
Catalyst (environmental concerns/consciousness), 35–37
Catalytic converter, 257
Catalytic reduction controls, 231
Causeway breaches, 179
CCC. *See* California Coastal Commission.
Celeron Corp., 195
Center for Marine Conservation, 12
Center for Plant Conservation, 48
CERES code, 314
CFC. *See* Chlorofluorocarbons.
Champlain gas pipeline project, 233
Channelview, Texas, styrene monomer/propylene oxide plant, 254
Chem Systems, Inc., 270
Chemical hazard, 45, 58, 64, 123, 239–240, 313
Chemical inventory, 276
Chemical Manufacturers Association, 273
Chemophobia, 58, 263
Chernobyl accident, 53
Cherry Point, Washington, refinery, 6, 251
Chevron Corp., 6, 107–108, 116, 194–196
Chilean grapes, 48, 62–65
Chlorofluorocarbons, 46, 123, 265–266
Chugach Mountains, 28
Chukchi Sea, 186

Citizens' watchdog panel, 225
Claims office (Exxon Valdez spill), 17
Clean Air Act, 82–84, 88–89, 204–206, 244
Clean air laws, 83–91, 147–148. *See also* Clean Air Act.
Clean Air Working Group, 91
Clean alternate fuels, 84
Clean coal plants, 86
Clean coal technology, 86, 147, 149
Clean fuels vehicles, 85, 96, 235
Clean Sites, 318
Clean Water Act, 82, 203–204
Clean (Exxon Valdez definition of), 26–27
Cleanup cost (Exxon Valdez), 12, 17, 19, 21, 30–31
Cleanup measures/force (Exxon Valdez), 26
Cleanup status (Exxon Valdez), 26–31
Clear-cutting (forestry), 100
CNG. *See* Compressed natural gas.
Coal demand, 146–157
Coal gasification, 150
Coal market, 147
Coal Oil Point field, 199
Coal reserves, 147
Coalbed methane, 155
Coal-fired power plants, 147
Coastal Zone Management Act, 188, 198
Coastal zone management plan consistency, 198
Cofiring, 150
Cogeneration, 138–139
Cogenerators, 135, 137
Cold water flushing, 22
Colorado, 253
Columbia Glacier, 1
Columbia River, 227
Colville River delta, 202
Combined cycle gas-fired power plants, 137
Combustion products, 47, 86
Command/control approach vs. market incentive, 87
Communications, 296–299
Community input, 303
Compensation for denial, 199
Competition (petroleum industry/environmentalist groups), 311–317
Competitive position (U.S. refiners), 239
Compost theory, 267
Compressed natural gas, 246–248, 257

INDEX

Hydrocarbon emissions, 34
Hydrocarbon potential, 15, 180–188
Hydrocracking, 255
Hydroelectric power, 143
Hydrofluoric acid, 281, 301
Hydrogen sulfide, 194–195
HyOx Inc., 253

I

Ice road, 173
Implementation plans, 148
In situ burning, 4–5, 9, 22, 25
Incident command system, 302
Incineration permits, 272
Incinerators, 272
Incremental capacity expansion, 237
Independent marketers, 259
Independent Petroleum Association of
    America, 108
Independent power producers, 135, 137–
    138
Independent producers, 108
Independent refiners, 257
Indochina, 34
Indonesia, 155
Industrial boilers/furnaces, 273
Inert/hydrocarbon gas venting, 231
Infrastructure, 167, 194, 247
Infrastructure infiltration (environmental-
    ists), 70
Innovative solutions (environmental prob-
    lems), 317–318
Institute for Resource Management, 319
Institute of Gas Technology, 124
Intellectual diversification, 126
Internal Revenue Service, 244
International Association of Drilling Con-
    tractors, 207
International Bird Rescue and Research
    Center, 9
International community, 79
International Tanker Owners Federation
    (Intertanko), 210
Interstate Natural Gas Association of Amer-
    ica, 151
Inter-industry networking, 106
Inupiat Eskimos, 175–176
Investor Responsibility Research Center,
    135

In-house environmental services, 105
Iowa, 92
IPAA. *See* Independent Petroleum Associa-
    tion of America.
Isomerization capacity, 256

J

J.R. Ewing syndrome, 287–288
Jackson, Sen. Henry, 41
Jago River well, 167
Japan, 161, 209
Jet fuel, 6
John D. Rockefeller/Standard Trust, 288
John Gray Institute of Lamar University,
    279
Johnston, Sen. Bennett, 15, 163
Jorgenson, Prof. Dale W., 307

K

Kaktovik Inupiat Corp., 167
Kenai Peninsula, 19
Kenai, Arkansas, refinery, 6
Kennedy, Robert, 282
Kentucky, 92, 189
Kern County, CA, 99
Knight Island, 9
Kodiak Island, 13
Kuparuk River field, 160, 165, 174–175

L

La Mirada, California, polymers plant, 64–
    65
Laboratory testing, 59
Land disposal: ban, 271; capacity, 273; haz-
    ardous waste, 271
Landfill: capacity, 264; cost, 266; opera-
    tions/practices, 123, 266–267; research,
    267
Latouche Island, 10
Lawsuits/claims (Exxon Valdez), 23
Lead, 84
League of Conservation Voters, 72
League of Women Voters, 57
Lease income, 177
Lease sale moratorium, 181
Lease sales, 183
Leasing/exploration-development/produc-
    tion, 183
Least-cost approach, 136

328

INDEX

Legal counsel, 300
Legal issues (oil spill), 23–25
Liability, 220, 226–227
Lightering operations, 4, 10–11
Limited exploration approach, 171
Limits to growth mentality, 306
Lindstedt-Siva, June, 30
Liquefied natural gas, 155
Liquefied petroleum gas, 246–248
Lisburne field, 175
Little Smith Island, 9
LNG. *See* Liquefied natural gas.
Logging industry, 311
Logistics, 8, 255
LOOP. *See* Louisiana Offshore Oil Port.
Los Angeles, California, refinery, 57
Louisiana State University Center for Energy Studies, 125
Louisiana State Univ. Institute for Environmental Studies, 125
Louisiana, 92, 189
Louisiana Offshore Oil Port, 217, 220
Love Canal, 55
Lower 48 onshore hydrocarbon potential, 188–190
Lower 48 production, 8
Low-sulfur, high-quality coal, 146
LPG. *See* Liquefied petroleum gas.
Lujan vs. National Wildlife Federation, 189
Lujan, Interior Sec. Manual, 20, 216

M

Mackenzie Delta, 165
MACT. *See* Maximum achievable control technology.
Magnetic field, 35
Mahoney, Richard, 127
Main Bay, 9
Malin, Clement, 261
Management philosophy, 296–299
March of Dimes, 72
Marine sanctuary, 184
Marine mammals, 12–13
Market incentive, 87
Market loss, 268
Market niche, 255
Market response (oil spill), 6–7
Marketers' concerns, 258–261
Marsh Creek anticline, 165

Maryland, 92
Massachusetts, 92
Materials handling, 207
Maxibarges, 21
Maximum achievable control technology, 85, 277
McClure, Sen. James, 163, 186–187, 214
Measure A, 300–301
Media relations, 14, 40, 54, 62–65, 168, 287–304
Mega Borg, 92, 209, 228
Methane, 205
Methanol, 84, 88, 95, 239–243, 249, 257
Methanol production cost, 241
Methanol suppliers, 241
Methanol vehicles, 241–242
Methyl tertiary butyl ether, 243, 250
Metzenbaum, Sen. Howard, 19
Metzger, Dr. H. Peter, 73
Micromanagement of waste materials, 201
Middle East oil supplies, 36
Milne Point field, 165, 167, 174–175
Minerals Management Service, 20, 160, 182–183
Mobil Oil Corp., 24, 281, 300–301
Mobile source pollution, 251
Model plant, 280–281
Modular power plants, 137
Monkeywrenching. *See* Ecoterrorism.
Monongahela River, 48
Monsanto Corp., 127, 264, 274
Montague Island, 10
Monterey Bay area, 184
Mooring systems, 11
Mooring terminals, 216
Morrow, Chm. Richard, 59
Mothers and Others for Pesticide Limits, 63
Mousse, 12, 22
MTBE. *See* Methyl tertiary butyl ether.
Multiple use management, 188
Municipal waste, 264
Murkowski, Sen. Frank, 10
Murray, Chm. Allen E., 16
Muskie, Sen. Ed, 40

N

NAAQS. *See* National ambient air quality standards.